Hooray for Division Facts!

by
Becky Daniel

illustrated by Judy Hierstein

Math Facts Wheel by Kenneth Holland

Cover by Judy Hierstein

Copyright © Good Apple, Inc., 1990

ISBN No. 0-86653-520-9

Printing No. 9876543

**GOOD APPLE, INC.
BOX 299
CARTHAGE, IL 62321**

The purchase of this book entitles the buyer to reproduce the student activity pages for classroom use only. Any other use requires the written permission of Good Apple, Inc.

All rights reserved. Printed in the United States of America.

Table of Contents

Dividing Zero 1	Computer Division 36
Dividing by One 2	Double-Cross Division 37
Dividing by Two 3	Search and Circle 38
Dividing by Three 4	Common Quotients 39
Division Line Design 5	What's My Sign? 40
Drawing Division Dice 6	Arrow Division 41
Matchup 7	Connect Them All! 42
Dividing by Four 8	Division Line Design 43
Practice Dividing by Four 9	Hide and Seek Problems 44
Dividing by Five 10	Color It Division 45
Division Design 11	Checking Division 46
Dividing by Six 12	Just Two Little Lines 47
Magic Division Squares 13	Just Three Lines! 48
Dividing by Seven 14	The Bake Sale 49
Division Daisies 15	Count Your Chickens 50
The Great Eight Race 16	All the Marbles 51
Charter Division 17	Break the Bank 52
Letter Division 18	Dot-to-Dot Division 53
Coloring Division 19	Hide-and-Go-Seek Division 54
Dividing by Nine 20	Signing Division 55
Division Families 21	Making Arrangements 56
About Face! 22	The Magic Quilt 57
Division Pairs 23	Calendar Magic 58
Dividing by Ten 24	One More Time, Please 59
Dividing by Eleven 25	Fingers on One Hand 60
Perfect Pairs 26	Pick a Number, Any Number 61
Division Twisters 27	Pre/Post Test 62
Dividing by Twelve 28	Graphing Your Progress 65
Division Dart Board 29	Division Bingo 66
Paths of Division 30	Division Bingo Card 67
Ring Three 31	Top Eighty 68
Division Towers 32	Quotients Maze 69
The Division Connection 33	Reproducible Facts Wheel 70
Heart-Shaped Division 34	Answer Key 72
Coded Division 35	Award Certificates 76

To the Teacher

There may be only one thing more boring than TEACHING children division facts—LEARNING the facts. However, mastering the division facts doesn't have to be a boring, tedious process, not if you use *Hooray for Division Facts!* The puzzles, mazes, codes, magic tricks, games, learning aids and math facts wheel in this book were carefully designed to teach the basic addition facts while entertaining and delighting your mathematicians. Every page is a different activity format so children never get bored.

It is highly recommended that you begin this book by giving each student the three-page division facts test found on pages 62, 63 and 64. Allow exactly five minutes for the students to answer as many of the division problems as they can. Record the initial scores. As the children are learning the facts, retest them weekly using the same three-page test and recording the scores each time. You may want the children to graph their own progress. (See graph on page 65.) The test can also be used as a division facts aid. Simply cut along the solid lines to create mini flash cards so children can review the division facts with which they are experiencing difficulty. Each child should also make his/her own division facts wheel. See reproducible patterns on pages 70 and 71.

The many activity sheets found in *Hooray for Division Facts!* were carefully designed and sequenced so that each page adds new facts and reviews those already learned. With the exception of several activities in the back of the book, no regrouping is necessary. If children master the basic facts, mathematics is enjoyable; however, if they do not master those facts, mathematics will soon become difficult and frustrating. Don't let one single student leave your classroom this year not knowing the division facts. It's fun and it's easy to sharpen math skills with the colossal collection of ideas found herein.

Using the Math Facts Wheel

The self-teaching math wheel is a fantastic teaching tool for grades two, three and four. Students using this simple device learn the basic math facts without consuming valuable class teaching time. The teacher is freed from forcing exercises upon students because students actually enjoy the math wheel. Learning division facts becomes fun. The embarrassment that sometimes accompanies the use of flash cards is eliminated. To use the wheel, the student first gives his answer to himself, then advances the wheel slightly and presto! the correct answer appears! If he was correct, seeing the correct answer will reinforce the student's memory or confidentially correct his error. Moving through the full range of basic division facts on the wheel, the student rapidly achieves readiness to perform the basic math functions.

This math tool has demonstrated its usefulness with special students who are having learning difficulties. Slow learners require continuing reinforcement and correction. The math wheel meets these needs. Continued practice day after day helps these students achieve using the wheel without the teacher's constant supervision or assistance.

Assembly of the Math Facts Wheel

1. Punch out the two die cut wheels.
2. Punch out the small answer box sections from the smaller (top) wheel.
3. Using a metal brad fastener, fasten the wheels together with the smaller wheel on top. Spread the brad tips behind the large (bottom) wheel.

Assembly of Teacher-Produced Wheels

1. Reproduce a copy of each of the components of the wheel for each of your students.
2. Glue these copies to oaktag or poster board.
3. Cut out, using heavy scissors or sharp knife. (Do not allow students to use knives.)
4. Cut out answer windows with X-acto knife. It is important to stay within the windows when cutting.
5. Assemble as with steps used on original wheel (see instructions above).

How to Use the Math Facts Wheel

1. Hold the assembled wheel in both hands arranging the perimeter numbers of both wheels so that they match up.
2. For example, for the division wheel, first move the top wheel so that the 2 printed on its edge is directly below the 4 on the edge of the bottom wheel. The student should then think "4 ÷ 2 = 2." Now move the top wheel slightly clockwise. The answer 2 should appear in the answer window to reinforce or correct the student.
3. Continue moving the top wheel until the number 2 is directly below the 6 on the bottom wheel. Give your answer. Move the top wheel slightly clockwise. The answer 3 will appear.
4. Continue in this fashion with the 2's going up to 72 ÷ 2. Then proceed with the 3's, applying it to each number on the bottom wheel, etc.

Note: For the division problems involving a quotient with a remainder, no answers will appear.

Dividing Zero

When you divide zero by any number, the quotient (answer) is always zero. Solve each division problem in the maze below. Then use a blue crayon to draw a path through the maze.

Bonus: Use a red crayon to draw yet another path through the maze.

Dividing by One

When you divide any number by one, the quotient (answer) is the number you divided. Example: $9 \div 1 = 9$

Draw a picture and write a division problem for each story problem below. The first one has been completed for you.

1. Eric has a total of two marbles. If he puts one marble in each of his pockets, how many pockets does Eric have?

   $2 \div 1 = 2$

2. Jessica has three candy bars. If she eats one candy bar each day, how long will her candy bars last?

3. Misty has five apples. If she puts one apple in each of her bags, how many bags does Misty have?

4. John has four baseballs. If he puts one baseball in each of his boxes, how many boxes does John have?

Bonus: Draw a picture to represent this division problem: ten divided by ten equals one.

Dividing by Two

In a division problem, the first number tells you how many objects are to be divided. The second number tells you how many objects are in each set. The quotient (answer) is the number of sets. Example: 6 ÷ 2 = 3   • •   • •   • •

Circle dots to show each division problem below.

1. 10 ÷ 2 = 5
   • •   • •   • •   • •   • •

2. 8 ÷ 2 =
   •   •   •   •   •   •   •   •

3. 12 ÷ 2 =
   •   •   •   •   •   •   •   •   •   •   •   •

4. 2 ÷ 2 =
   •   •

5. 8 ÷ 1 =
   •   •   •   •   •   •   •   •

6. 10 ÷ 5 =
   •   •   •   •   •   •   •   •   •   •

7. 5 ÷ 1 =
   •   •   •   •   •

8. 8 ÷ 4 =
   •   •   •   •   •   •   •   •

Bonus: Draw dots and circle sets to show these division problems: 20 ÷ 10 and 20 ÷ 2.

Dividing by Three

Draw dots and circles to show each division problem below. Then write the quotient for each problem. Remember, the first number tells you the total number of dots. The second number tells you the number in each set and the quotient tells you the number of sets.

1. $3 \div 3 =$

2. $9 \div 3 =$

3. $10 \div 2 =$

4. $10 \div 5 =$

5. $18 \div 3 =$

6. $18 \div 6 =$

$6 \div 3 =$

$12 \div 4 =$

$12 \div 3 =$

$6 \div 2 =$

$21 \div 3 =$

$0 \div 3 =$

Bonus: In the division problem $100 \div 5$, how many sets are there?

Division Line Design

Color spaces with a quotient of 2 RED. Color spaces with a quotient of 3 YELLOW. Color all other spaces BLUE.

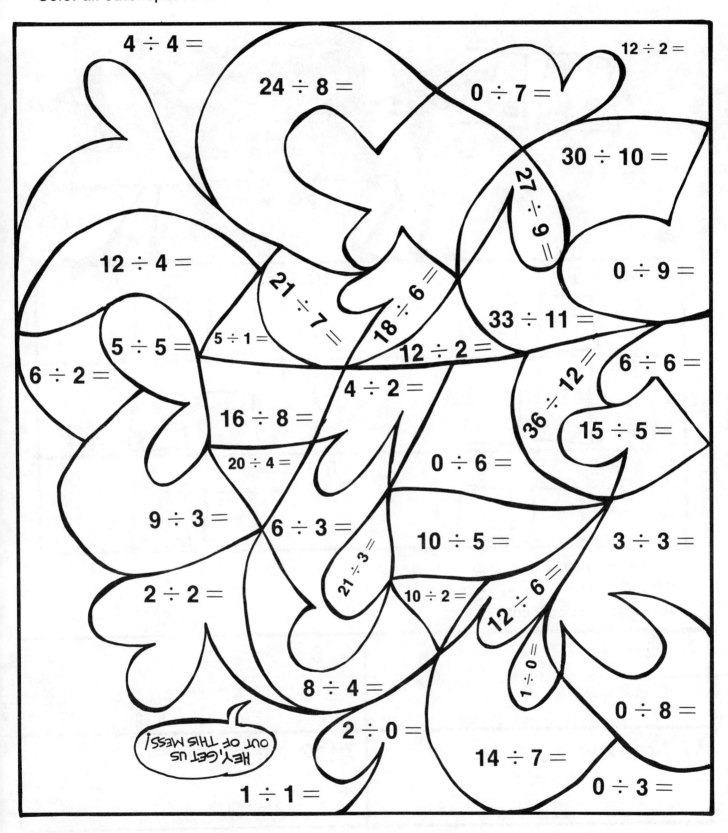

Bonus: List four numbers that can evenly be divided by both 2 and 3.

Drawing Division Dice

Write each quotient and then draw dots on the sets of dice below to show each division problem. Example: 4 ÷ 2 = 2

1. 12 ÷ 6 =   10 ÷ 5 =

2. 2 ÷ 1 =   6 ÷ 3 =

3. 4 ÷ 2 =   8 ÷ 4 =

Draw as many dice as needed to show each division problem below.
Example: 6 ÷ 2 = 3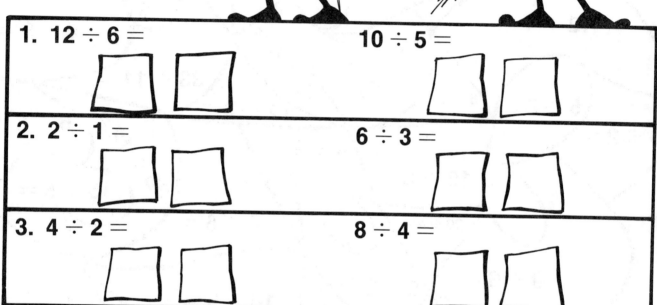
Then write each quotient.

4. 12 ÷ 1 =   5. 6 ÷ 2 =

6. 12 ÷ 2 =   7. 12 ÷ 4 =

Bonus: Draw dice to show the division problem 30 ÷ 5.

Matchup

To discover the best way to learn division facts, draw a line connecting division problems that have the same quotient. Then fill in the boxes with the letters intersected by a line.

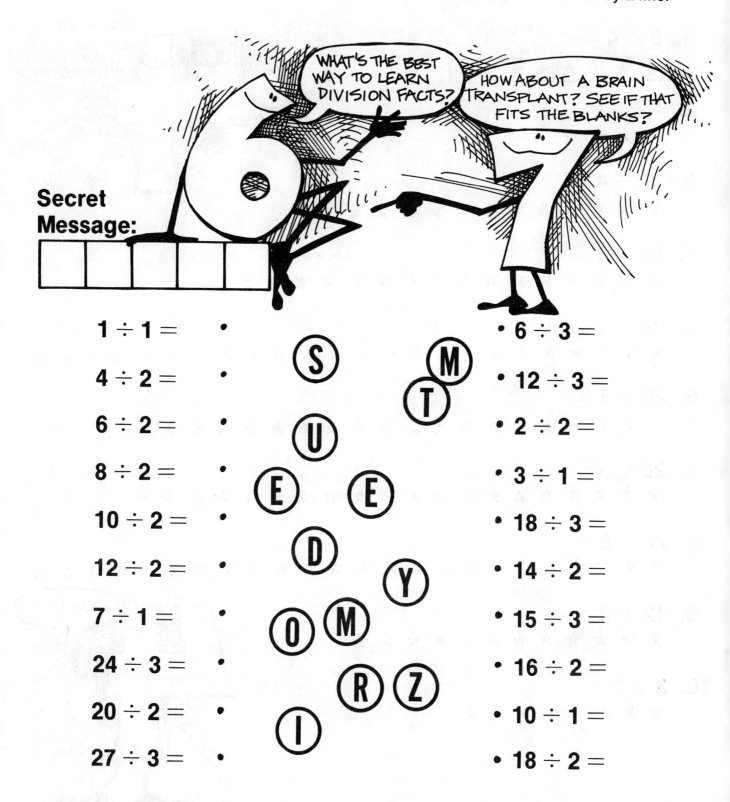

Secret Message:

☐ ☐ ☐ ☐ ☐

1 ÷ 1 =	6 ÷ 3 =
4 ÷ 2 =	12 ÷ 3 =
6 ÷ 2 =	2 ÷ 2 =
8 ÷ 2 =	3 ÷ 1 =
10 ÷ 2 =	18 ÷ 3 =
12 ÷ 2 =	14 ÷ 2 =
7 ÷ 1 =	15 ÷ 3 =
24 ÷ 3 =	16 ÷ 2 =
20 ÷ 2 =	10 ÷ 1 =
27 ÷ 3 =	18 ÷ 2 =

Letters between columns: S, M, T, U, E, E, D, Y, O, M, R, Z, I

Bonus: To discover yet another way to learn division facts, unscramble the letters that are not intersected by a line.

Dividing by Four

Circle the stars to show each division problem below. Remember, the number of sets is the quotient. The first one has been completed for you.

1. 8 ÷ 4 =
   [★ ★ ★ ★] [★ ★ ★ ★]

2. 4 ÷ 4 =
   ★ ★ ★ ★

3. 12 ÷ 4 =
   ★ ★ ★ ★ ★ ★ ★ ★ ★ ★ ★ ★

4. 16 ÷ 4 =
   ★ ★ ★ ★ ★ ★ ★ ★ ★ ★ ★ ★ ★ ★ ★ ★

5. 24 ÷ 4 =
   ★

6. 20 ÷ 4 =
   ★

7. 20 ÷ 5 =
   ★

8. 24 ÷ 6 =
   ★

9. 12 ÷ 3 =
   ★ ★ ★ ★ ★ ★ ★ ★ ★ ★ ★ ★

10. 8 ÷ 2 =
    ★ ★ ★ ★ ★ ★ ★ ★

Bonus: Draw stars and circle them to prove that 48 ÷ 4 = 12.

Practice Dividing by Four

Example: ⊐∪ ÷ ⊏ = 42 ÷ 6 = 7

Using the box code, complete each division problem below.

1. ⊐⊓ ÷ ⊐ =   ∪⊓ ÷ ⊐ =   ⌊∪ ÷ ⊐ =

2. ∪⊐ ÷ ⊏ =   ⌊⊏ ÷ ⊐ =   ⊐⊐ ÷ ⊐ =

3. ⌡⊏ ÷ ⊐ =   ⌡∪ ÷ ⌊ =   ⊐ ÷ ⌡ =

4. ⌊⊏ ÷ ⌈ =   ∪⊐ ÷ ⊐ =   ⊓ ÷ ∪ =

5. ⌡∪ ÷ ⊐ =   ⊓ ÷ ⊐ =   ⊐ ÷ ⊐ =

6. ∪⊓ ÷ ⌈ =   ⊐⊓ ÷ ⌡∪ =   ⌊∪ ÷ ⊓ =

Bonus: If zero equals X, write four more division problems in code.

Dividing by Five

Circle every fifth number. The first row has been done for you. Then use the circled numbers to solve the division problems below.

1	2	3	4	⑤	6	7	8	9	⑩
11	12	13	14	15	16	17	18	19	20
21	22	23	24	25	26	27	28	29	30
31	32	33	34	35	36	37	38	39	40
41	42	43	44	45	46	47	48	49	50
51	52	53	54	55	56	57	58	59	60

1. $25 \div 5 =$ $35 \div 5 =$
2. $30 \div 5 =$ $10 \div 5 =$
3. $60 \div 5 =$ $40 \div 5 =$
4. $45 \div 5 =$ $50 \div 5 =$
5. $15 \div 5 =$ $20 \div 5 =$
6. $5 \div 5 =$ $55 \div 5 =$

HANDS AND FEET HAVE ALWAYS MADE MY LIFE SO EASY – 1, 2, 3, 4, 5

Bonus: What is the smallest number that can evenly be divided by 1, 2, 4, 5 and 10?

Division Design

Complete each division problem in the line design below. Color sections of the design with a quotient of 2 ORANGE. Color sections of the design with a quotient of 3 PURPLE. Color sections of the design with a quotient of 4 GREEN. Color sections of the design with a quotient of 5 BLUE.

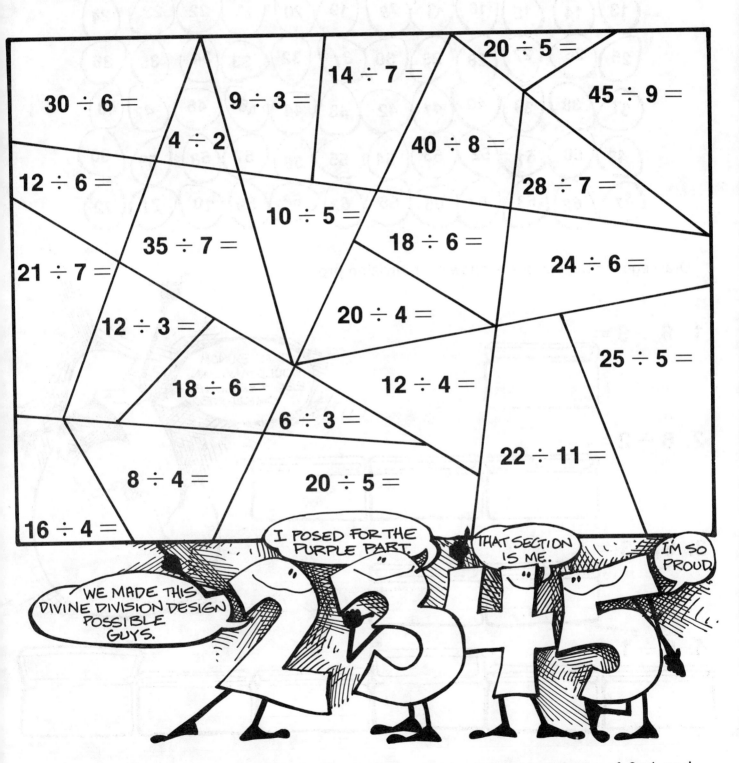

Bonus: Draw your own division design. Include problems with quotients of 0, 1 and 2. Ask a friend to color your division line design.

Dividing by Six

Color every sixth egg YELLOW.

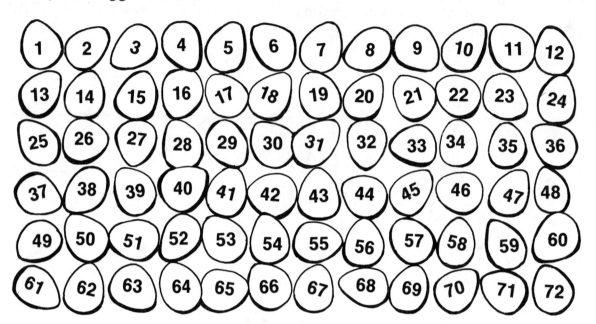

Draw eggs in the cartons to show each division problem.

1. 6 ÷ 6 =

2. 6 ÷ 2 =

3. 6 ÷ 3 =

4. 6 ÷ 1 =

Bonus: Which equals more, ten dozen divided by half a dozen or five dozen divided by ten?

Magic Division Squares

Complete each magic square. The last number in each row and each column must be the quotient for the first and second numbers in that row or column. Study the example that has been completed for you.

Bonus: Make up your own division magic square.

Dividing by Seven

Complete the number chart below. Then use the last number in each row to solve the division problems at the bottom of the page.

1	2	3	4	5	6	7
8						
15	16					
22	23	24				
29	30	31	32			
36	37	38	39	40		
43	44	45	46	47	48	49
50	51	52	53	54		
57	58	59	60			
64	65	66				
71	72					
78						

1. $21 \div 7 =$
2. $84 \div 7 =$
3. $77 \div 7 =$
4. $35 \div 7 =$
5. $56 \div 7 =$
6. $42 \div 7 =$

$70 \div 7 =$
$49 \div 7 =$
$63 \div 7 =$
$28 \div 7 =$
$7 \div 7 =$
$14 \div 7 =$

Division Daisies

To complete the division daisies, divide the number in each petal by the number in the center. Example $32 \div 4 = 8$

Bonus: Complete a division daisy with the number 6 in the center.

The Great Eight Race

Time yourself. How many minutes will it take you to race around the track? Begin in the space after GO and call out the answers to a friend. Can you complete the race in less than three minutes? Two minutes? Practice until you can do it in less than one minute.

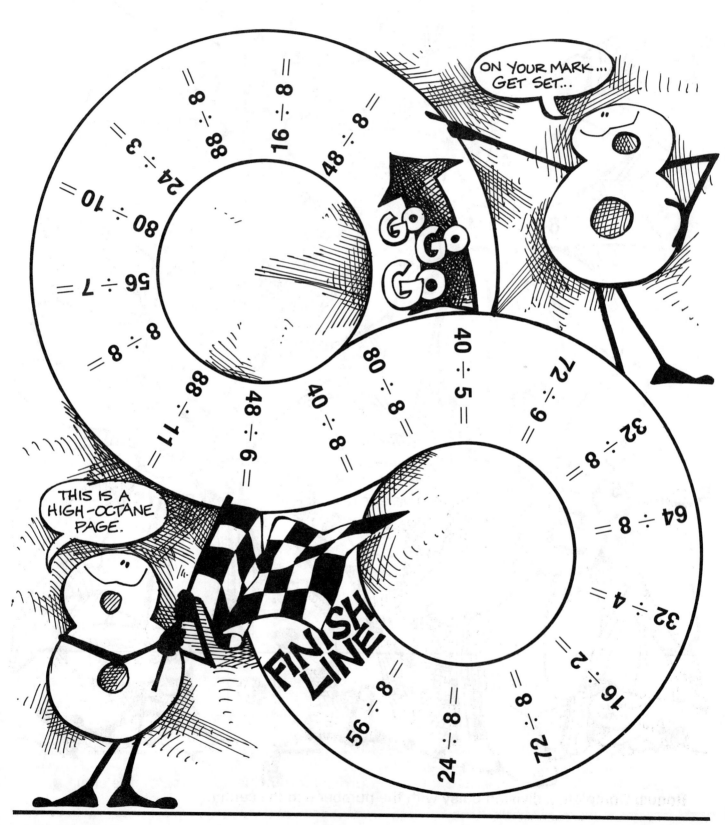

Charter Division

Use the chart below to list and solve each division problem indicated by dots. The first row has a dot under the numbers 4 and 1, so the problem is 4 ÷ 1.

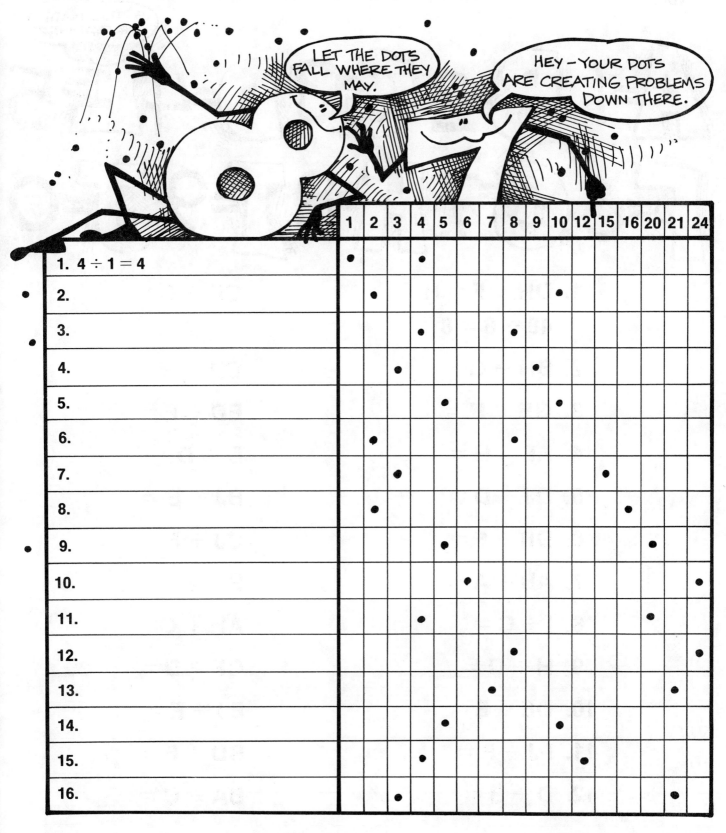

Bonus: Using the chart above, list three new division problems not found on the chart.

Letter Division

Using the letter code, solve each problem below. The first one has been completed for you.

A=1 B=2 C=3 D=4 E=5
F=6 G=7 H=8 I=9 J=0

1. DH ÷ F = H
   48 ÷ 6 = 8

2. BH ÷ G =

3. GB ÷ H =

4. AF ÷ H =

5. DJ ÷ D =

6. DB ÷ F =

7. AB ÷ A =

8. I ÷ C =

9. H ÷ D =

10. DE ÷ E =

11. FJ ÷ F =

12. D ÷ B =

CE ÷ G =

CB ÷ H =

ED ÷ F =

D ÷ D =

BJ ÷ E =

CJ ÷ F =

B ÷ B =

AH ÷ C =

CF ÷ D =

EJ ÷ E =

BD ÷ F =

BA ÷ C =

Bonus: Write three new division problems in code. Write the quotients in code, too.

Coloring Division

Solve each division problem in the spaces below. Color spaces with a quotient of 5 BLUE. Color spaces with a quotient of 6 YELLOW. Color spaces with a quotient of 7 GREEN.

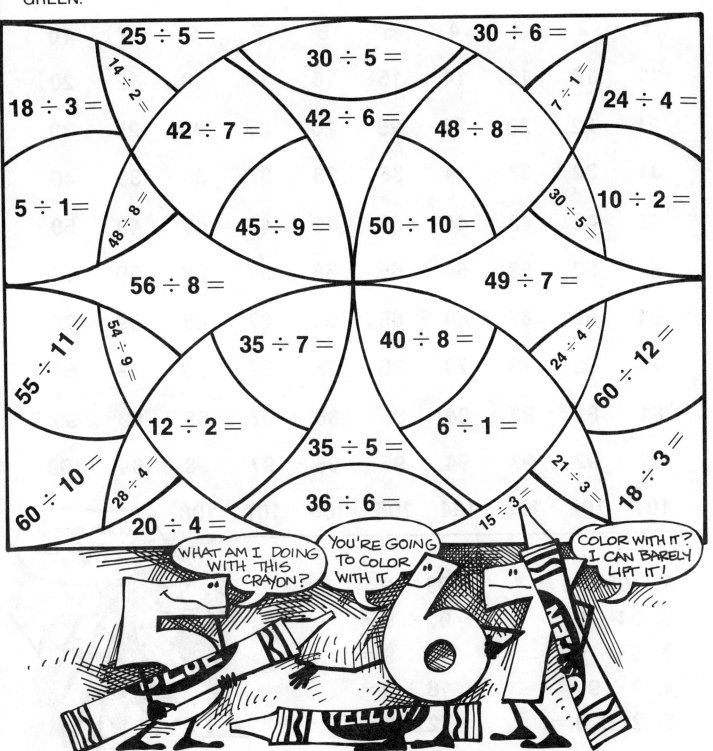

Bonus: 1. What is the smallest number that can evenly be divided by 5 and 7 but not by 6? 2. What is the smallest number that can evenly be divided by 5 and 6 but not by 7? 3. What is the smallest number that can evenly be divided by 6 and 7 but not by 5?

Dividing by Nine

Circle every ninth number. Use the circled numbers to solve the division problems below.

1	2	3	4	5	6	7	8	⑨	10
11	12	13	14	15	16	17	⑱	19	20
21	22	23	24	25	26	27	28	29	30
31	32	33	34	35	36	37	38	39	40
41	42	43	44	45	46	47	48	49	50
51	52	53	54	55	56	57	58	59	60
61	62	63	64	65	66	67	68	69	70
71	72	73	74	75	76	77	78	79	80
81	82	83	84	85	86	87	88	89	90
91	92	93	94	95	96	97	98	99	100
101	102	103	104	105	106	107	108		

1. 108 ÷ 9 =   99 ÷ 9 =
2. 54 ÷ 9 =   63 ÷ 9 =
3. 36 ÷ 9 =   45 ÷ 9 =
4. 9 ÷ 9 =   18 ÷ 9 =
5. 27 ÷ 9 =   72 ÷ 9 =

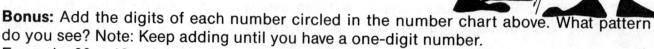

Bonus: Add the digits of each number circled in the number chart above. What pattern do you see? Note: Keep adding until you have a one-digit number.
Example: 99 = 18 = 1 + 8 = 9

Division Families

Complete each box below by listing twelve pairs of numbers with a quotient indicated by the large number found in the corner of each box. A few have been listed for you in the first box.

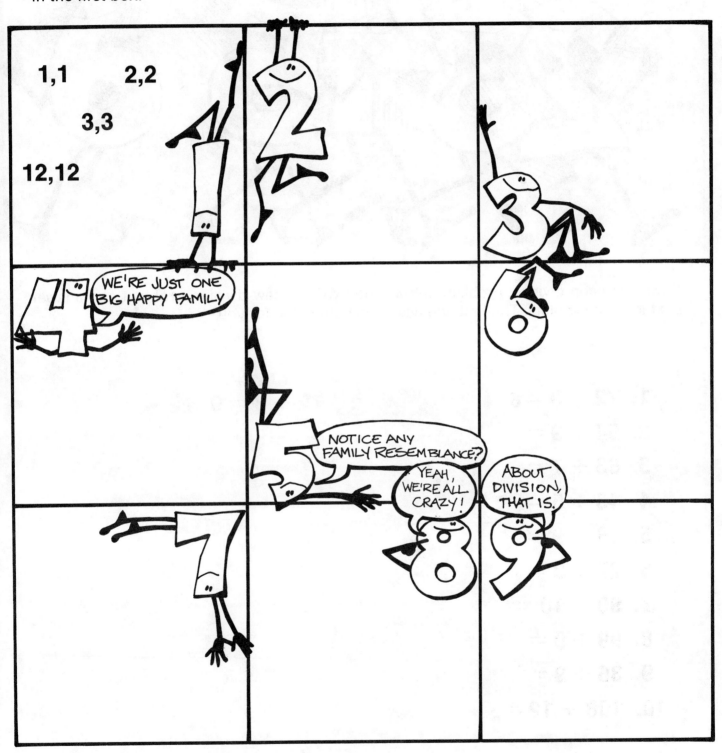

Bonus: How many different numbers can you write using the three digits 1, 2 and 3? You do not have to use all three digits in each number, but you may not repeat a digit in any number. Example: 13 is OK, but 33 is not a correct answer.

About Face!

Complete each division problem below. Then write a new division problem and answer next to each problem. The first one has been completed for you.

1. $72 \div 9 = 8$ $72 \div 8 = 9$
2. $54 \div 9 =$
3. $63 \div 7 =$
4. $45 \div 5 =$
5. $18 \div 9 =$
6. $27 \div 3 =$
7. $90 \div 10 =$
8. $99 \div 9 =$
9. $36 \div 9 =$
10. $108 \div 12 =$

Bonus: What two-digit number can evenly be divided by 9 but cannot be written two different ways?

Division Pairs

Write a number pair in each circle connected by a line that has a quotient indicated by the number in the center of each design. The first one has been started for you.

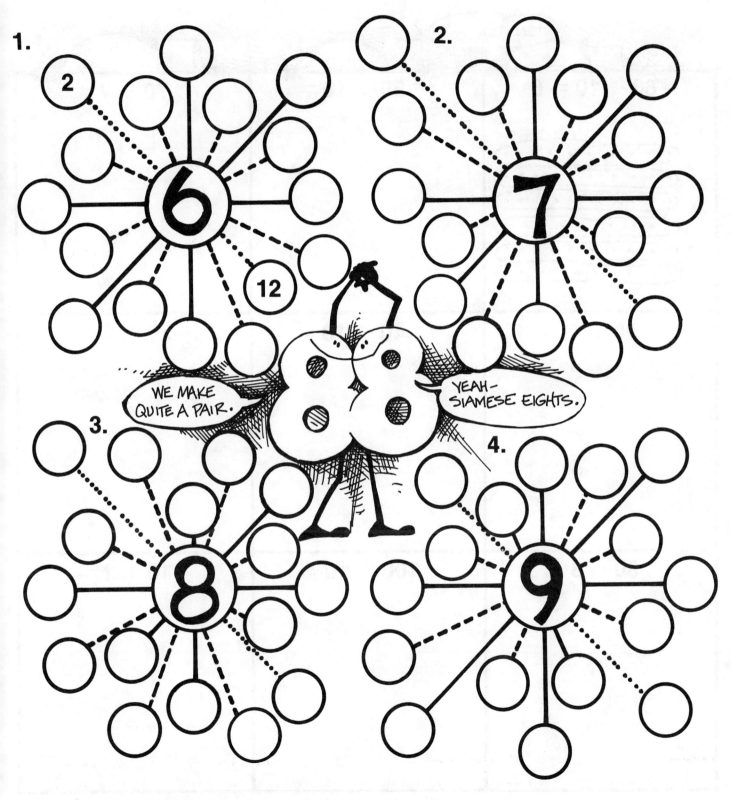

Bonus: Draw a design with the quotient 12 in the center.

Dividing by Tens

Draw and circle dots in each box to prove the division problems below. Remember the first number tells the total number of dots. The second number tells how many are in each set, and the quotient tells how many sets.

TEN WAS HERE

60 ÷ 10 = 6	50 ÷ 10 =	70 ÷ 7 =
90 ÷ 10 =	40 ÷ 10 =	20 ÷ 2 =
80 ÷ 8 =	100 ÷ 10 =	10 ÷ 1 =

Bonus: Circle dots to prove that 120 ÷ 12 = 10.

Dividing by Eleven

Dividing by 11 is very easy. Complete the number chart below. Using the last number in each row, solve each division problem.

1	2	3	4	5	6	7	8	9	10	11
12	13	14	15	16	17	18	19	20	21	22
23	24	25	26	27						
34	35	36	37							
45	46	47								
56	57									
67										
78	79									
89	90	91								
100										

1. $99 \div 11 =$
2. $11 \div 11 =$
3. $110 \div 10 =$
4. $88 \div 8 =$
5. $77 \div 11 =$
6. $55 \div 5 =$
7. $44 \div 4 =$
8. $77 \div 7 =$

$110 \div 11 =$
$33 \div 11 =$
$99 \div 9 =$
$66 \div 11 =$
$66 \div 6 =$
$22 \div 2 =$
$121 \div 11 =$
$132 \div 11 =$

Bonus: Except for the very last number on your chart, what pattern do you see when you look at the last number in each row above?

Perfect Pairs

Circle all the number pairs that have a quotient of 10 with a blue crayon. Circle all the number pairs that have a quotient of 11 with a red crayon.

10	3	10	33	4	6	12	3	9	90	6	20
1	10	8	1	10	7	9	1	88	40	9	2
7	9	70	55	5	8	60	0	8	5	8	90
1	6	7	60	3	90	7	5	48	10	1	3
10	66	8	80	6	4	55	5	10	1	7	60
70	6	1	8	33	0	9	8	50	7	5	4
7	0	10	11	4	3	1	1	20	2	11	1
40	44	99	1	11	0	0	2	13	12	0	11
4	4	9	10	1	0	44	0	30	3	88	1
55	5	77	10	1	1	4	0	0	7	8	10
4	5	7	1	9	4	3	0	77	0	0	1
50	5	8	66	6	99	9	1	3	0	10	0

Bonus: Draw your own number maze for the quotient 9.

Division Twisters

Read each tongue twister below. Then write and solve each division problem.

1. Six slithering, slimy snails sniffed sixty-six savory sausages and proceeded to slurp them all. If equally divided, how many savory sausages did each slithering, slimy snail slurp?

2. Five famous frogs flipped five times when they saw fifty-five frosty freezes. If the famous frogs sipped equally the fifty-five frosty freezes, how many frosty freezes did each famous frog sip?

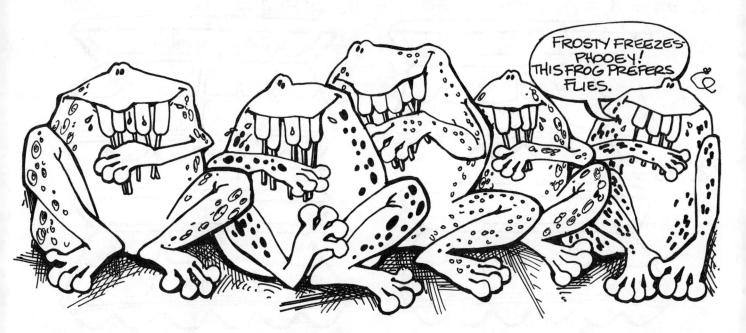

3. Eight elegant elephants ate eighty-eight great, green grapes. How many great, green grapes did each elegant elephant eat?

4. Four fabulous fish found forty-four foaming, frothing figs at the bottom of the pond. If the four fabulous fish equally divided the foaming, frothing figs, how many did each fabulous fish consume?

5. Nine nanny goats knitted a total of ninety-nine nifty napkins. If the nine nannys knitted an equal number of nifty napkins, how many nifty napkins did each nanny goat knit?

Bonus: Thirty-three thirsty thrushes each flew three miles to get a drink. How many miles did each thirsty thrush fly?

Dividing by Twelve

Place an equal number of dots in each section of the egg cartons to show the division problems. Write the answers, too.

1. 24 ÷ 2 =
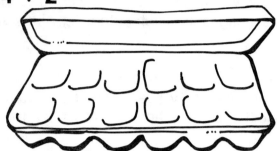

5. 60 ÷ 5 =
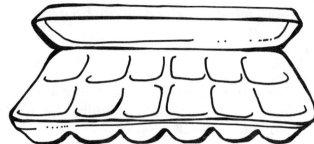

2. 96 ÷ 8 =
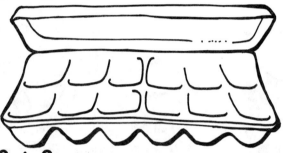

6. 48 ÷ 4 =
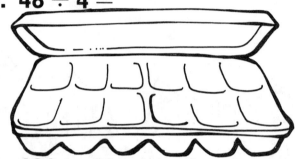

3. 36 ÷ 3 =
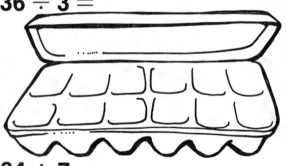

7. 132 ÷ 11 =
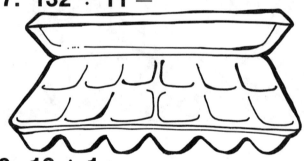

4. 84 ÷ 7 =
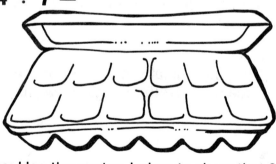

8. 12 ÷ 1 =

Bonus: Use the carton below to show that 36 ÷ 6 = 6.
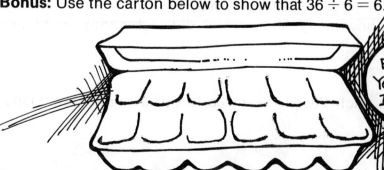

Division Dart Board

Divide the number in the center ring by the numbers in the inner ring, and write the quotients in the outer ring.

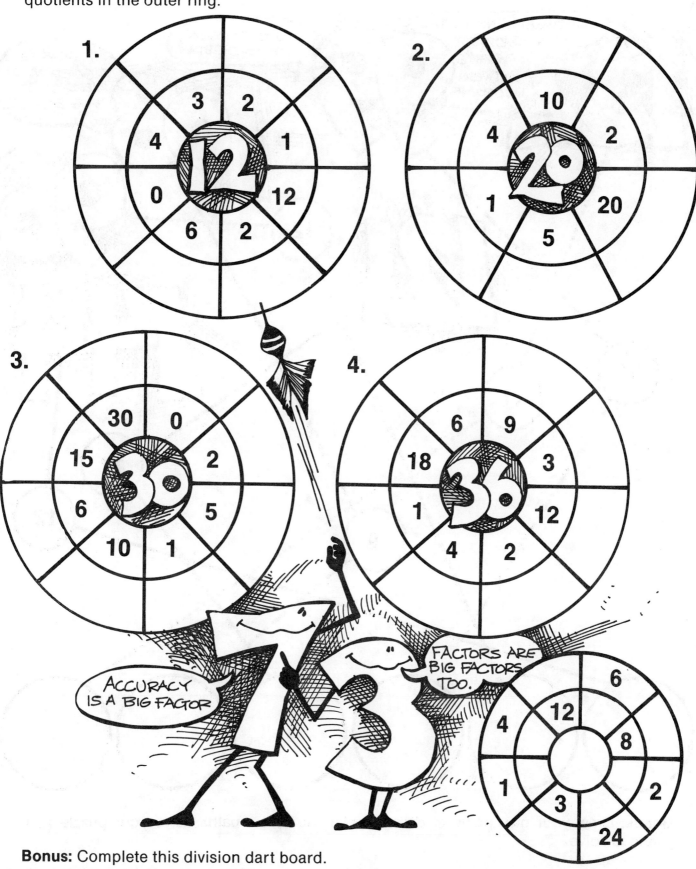

Bonus: Complete this division dart board.

Paths of Division

Follow each path below, dividing the first number you see by the second number found on the path. Write the quotient in the circle at the end of each path.

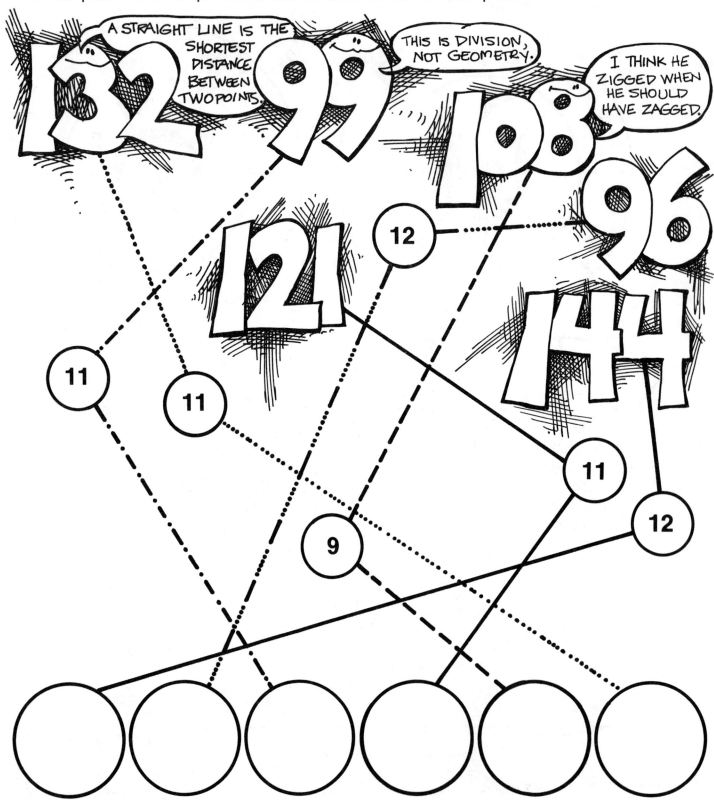

Bonus: Draw your own paths of division with five or six paths. Give your puzzle to a friend to solve.

Ring Three

Find and circle exactly ten three-number combinations that are true division sentences in each box below. When you complete the puzzles, every number will be circled once and only once. In the first box, some of the division sentences have been circled for you.

```
(144 ÷ 12  =  12)   132    12    11
 (1  ÷  1  =   1)     2     1     2
   56    63    7      9    84     8
    7    70    7     10     7     8
    8    77    7     11    12     1
  120    12   10    108    12     9
```

```
  120   12   10   108   12    9
   96   84   12     7   72   60
   12   24   12     2   12   12
    8   12   12     1    6    5
   48   12    4    36   12    3
```

```
    5    1    5    33   55   60
   24    3    8     3    5   12
   27   30   36    11   11    5
    3    3    3    44   11    4
    9   10   12    48    4   12
```

```
    3    2    2     1    4   14
    1   12    2     6    4    2
    3    8    2     4    1    7
    4    1    4     8    2    4
   10    2    5    12    2    6
```

Bonus: Can you solve this very tricky ring puzzle?

```
0 6 0 3 0 3
3 6 2 3 1 3
0 1 1 1 1 1
2 1 2 3 1 3
2 0 0 0 0 0
1 0 0 6 0 0
```

Division Towers

Divide each number pair connected by a line and place the quotient in the circle at the top. Work your way to the top of each tower. Some answers have been filled in for you in the first tower.

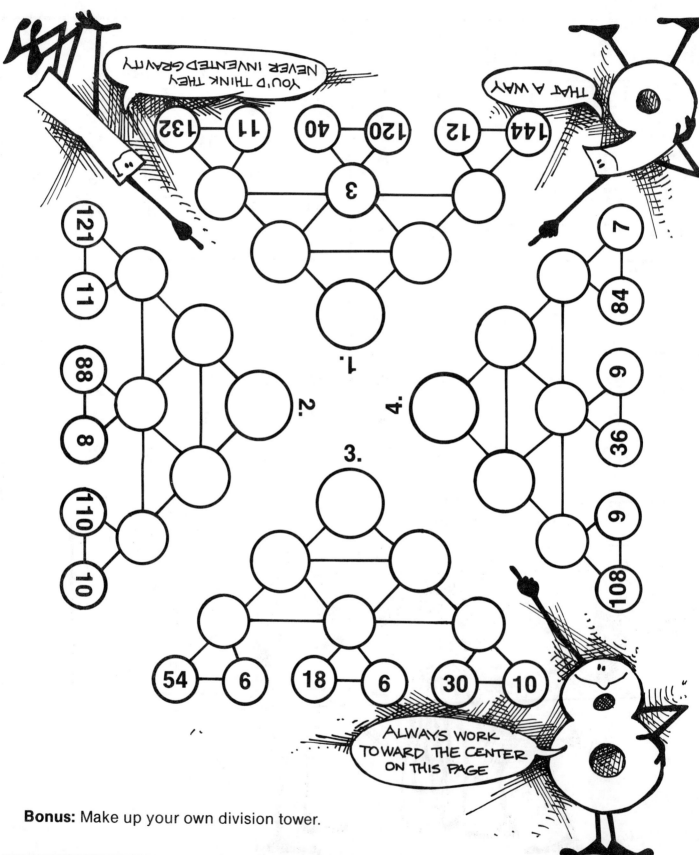

Bonus: Make up your own division tower.

The Division Connection

To discover the only fair way for two hungry children to divide a small piece of pie, solve each division problem below; then place the appropriate letter in each box.

Secret Message:

☐☐☐ ☐☐☐☐ ,
1 2 3   4 5 6 7

☐☐☐ ☐☐☐☐☐
6 8 3   1 6 8 3 9

☐☐☐☐☐☐☐   ☐☐☐   ☐☐☐☐☐ .
4 8 1 1 7 3 7   8 10 7   11 10 3 4 3

1. 16 ÷ 4 = (c)	6. 11 ÷ 11 = (o)	9. 49 ÷ 7 = (s)
2. 9 ÷ 3 = (e)	7. 99 ÷ 9 = (p)	10. 36 ÷ 6 = (t)
3. 88 ÷ 11 = (h)	8. 63 ÷ 7 = (r)	11. 25 ÷ 5 = (u)
4. 110 ÷ 11 = (i)	 I'M SO HUNGRY MY STOMACH FEELS HOLLOW	12. 144 ÷ 12 = (a)
5. 12 ÷ 6 = (n)		13. 0 ÷ 9 = (m)

Bonus: To discover what kind of pie the children wanted to divide, fill in the appropriate letters.

☐☐☐☐☐☐☐   ☐☐☐☐☐
4 1 4 1 2 5 6   4 9 3 12 0

Heart-Shaped Division

Bonus: If a circle equals the number zero, write your telephone number in the heart-shaped code.

Coded Division

Use the code to solve each division problem.

1 = ○	2 = ⊘	3 = □	4 = ⊠	5 = ⊗
6 = ▱	7 = △	8 = ⊛	9 = ⊟	0 = △̷

1. ○⊘△ ÷ ○○ =   ⊠⊛ ÷ ○○ =

2. ○⊗ ÷ ⊗ =   ⊘△ ÷ ⊗ =

3. ▱ ÷ □ =   □▱ ÷ ○⊘ =

4. ○⊘ ÷ ▱ =   ○⊘ ÷ ○⊘ =

5. ⊘⊠ ÷ ⊠ =   ⊘⊠ ÷ ○⊘ =

6. □△ ÷ ▱ =   ⊘⊗ ÷ ⊗ =

7. ▱⊠ ÷ ⊛ =   ○⊛ ÷ ▱ =

8. ⊘⊠ ÷ ⊛ =   △⊘ ÷ ⊛ =

Bonus: Write ten new division problems in code. Write the quotients in code, too.

Computer Division

Using each number below, complete each division problem using the flow chart. Example: The first number is sixteen. Sixteen is an even number, so we divide by two (16 ÷ 2 = 8). Follow the arrow and add nine to eight (9 + 8 = 17). Seventeen is the answer to the first problem.

1. 16
2. 10
3. 15
4. 35
5. 14
6. 45
7. 22
8. 24
9. 5
10. 55
11. 20
12. 8

Bonus: What odd number used as input will have the same answer as using the even number 10 as input?

Double-Cross Division

Example: 1 = ⟩ and 5 = ⟨•

Use the code to solve each division problem below.

1. ⟩⌄⌄ ÷ ⟩∧ = ⟩∧
   144 ÷ 12 = 12

2. ⟨⌄ ÷ ∧ =

3. ∧⌄ ÷ ⟩∧ =

4. ⟨⟨• ÷ ⟩⟩ =

5. ∧⌄ ÷ ⟩∧ =

6. ⟩∧ ÷ ∧ =

7. ⟩ ÷ ⟩ =

8. ⟨∧ ÷ ⌄ =

⟨•̇ ÷ ⟩∧ =

⟩∧ ÷ ⟩∧ =

⟨⌄ ÷ ⟩⌄ =

⟩⟨∧ ÷ ⟩∧ =

⌄⌄ ÷ ⟨• =

⟨•̇ ÷ ⟩∧ =

∧∧ ÷ ⟩⟩ =

⟩⌄ ÷ ∧ =

Bonus: In code, write 100 ÷ 10 = 10 and 121 ÷ 11 = 11.

Search and Circle

Find and circle twenty-five true division sentences in the number maze below. Division sentences may be written across, down or diagonally. One has been circled for you.

Bonus: Find and circle another fifteen true division sentences hidden in the number maze.

Common Quotients

Look at the numbers in each box below. Decide what is the smallest number that can evenly be divided into each number? Write the number in the top left-hand square of each box.

1. 240 27 36 33	**2.** 77 14 84 49	**3.** 10 1110 100 110	**4.** 40 28 44 12
5. 36 132 144 24	**6.** 105 60 75 15 120	**7.** 121 33 99 11	**8.** 72 54 108 81

Bonus: What number can evenly be divided into each of these numbers: 86, 129, 215 and 301?

What's My Sign?

If you put a division sign and an equal sign in the right place in each group of digits below, you will make a true division sentence. The first one has been completed for you.

1. 4 8 ÷ 6 = 8
2. 8 4 1 2 7
3. 6 0 6 1 0
4. 6 3 7 9
5. 1 2 0 1 2 1 0
6. 3 5 7 5
7. 4 9 7 7
8. 6 4 8 8
9. 4 8 1 2 4
10. 7 2 8 9
11. 2 4 1 2 2
12. 5 6 8 7
13. 1 4 4 1 2 1 2
14. 2 4 6 4
15. 4 2 6 7
16. 1 2 1 1 1 1 1

6 0 1 2 5
5 4 6 9
9 6 1 2 8
7 0 7 1 0
2 8 7 4
1 3 2 1 2 1 1
7 2 1 2 6
3 6 1 2 3
4 2 7 6
1 3 2 1 1 1 2
1 2 0 1 2 1 0
5 6 7 8
4 8 8 6
1 8 6 3
1 2 0 1 0 1 2
1 1 0 1 1 1 0

Bonus: Make up your own puzzle like this one, but the numbers used must be a pattern. Example: 1 2 1 2 1 (12 ÷ 12 = 1)

Arrow Division

Use the chart below to complete the division problems. What looks like a secret code is actually just division problems. The arrow tells you what direction to look on the chart to find the number to divide the given number by. Example: 18 = 18 ÷ 3. The arrow stands for the number directly below the given number 18.

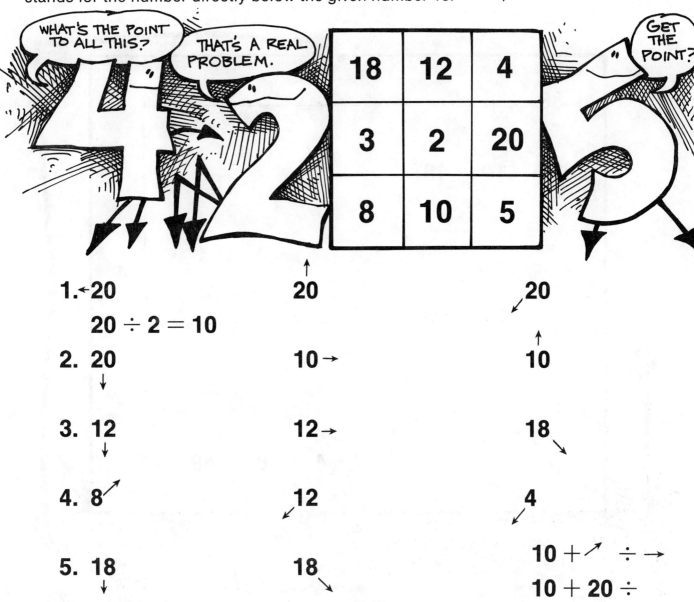

Bonus: Arrange the digits 2, 3, 4, 5, 8, 10, 12, 18 and 20 in the design. Then list five new problems and answers with your new arrow code.

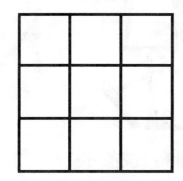

1.
2.
3.
4.
5.

Connect Them All!

Draw lines to connect the number pairs with a quotient of 6. You may not cross any lines. You must connect every dot with another dot. Example: The numbers 36 and 6 have been connected for you because 36 ÷ 6 = 6.

Bonus: Create your own connect-the-number pairs with a quotient of 7.

Division Line Design

Using a ruler, draw a line connecting each number divisible by 3 to each star. Example: Connect the number 9 to all the stars because $9 \div 3 = 3$.

Bonus: Color areas of your line design to create a pattern.

Hide and Seek Problems

On the lines under each circle, list two different division problems using all the digits inside the circle. The first one has been completed for you.

1.

   99 ÷ 9 = 11
   99 ÷ 11 = 9

2.

3.

Bonus: Can you write a true division sentence using these numbers: 0, 0, 0, 0, 1?

Color It Division

Complete the division problems in the picture. Color areas with a quotient of 8 BLUE. Color areas with a quotient of 9 RED. Color areas with a quotient of 12 YELLOW.

Bonus: 1. What is the smallest number that can evenly be divided by 8 and 12 but not by 9? 2. What is the smallest number that can evenly be divided by 9 and 12 but not by 8?

Checking Division

Place two checks in each row to create a true division problem. The first one has been done for you.

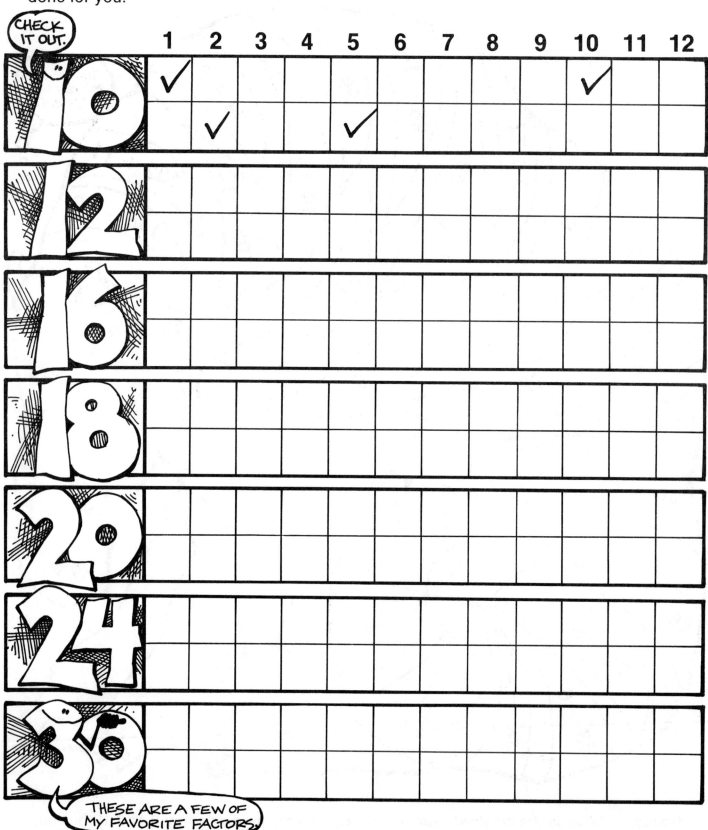

Bonus: List a number that can be shown three different ways on the check chart.

Just Two Little Lines

Draw two straight lines to divide the pizza so that each area has two numbers with a quotient of 6.

Draw two straight lines to divide the pizza so that each area has two numbers with a quotient of 7.

Draw two straight lines to divide the pizza so that each area has two numbers with a quotient of 8.

Bonus: Draw two straight lines to divide the pizza so that each area has two numbers with the same quotient.

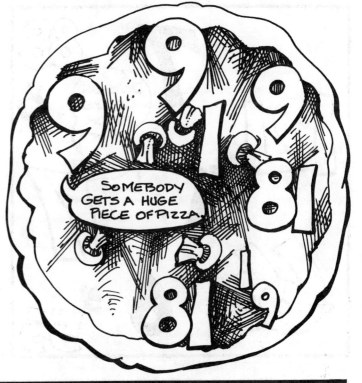

Just Three Lines!

Draw three straight lines to divide the square so that each area has two numbers with the quotient of 4.

Draw three straight lines to divide the square so that each area has two numbers with the quotient of 5.

Draw three straight lines to divide the square so that each area has two numbers with the quotient of 11.

Bonus: Draw three straight lines to divide the square so that each area has two numbers with the same quotient.

The Bake Sale

Write and solve the division problem for each story problem below.

1. Jenny has 64¢. How many peanut butter cookies can she buy?

2. Nancy has $1.00. How many chocolate chip cookies can she buy?

3. Max has $1.44. How many fortune cookies can he buy?

4. Mike has 72¢. How many oatmeal raisin cookies can he buy?

5. Charlie has twice as much money as Mike. How many peanut butter cookies can he buy?

6. How many fortune cookies can Charlie buy?

7. Lisa has $1.32. How many fortune cookies can she buy?

8. Crystal has six quarters. How many chocolate chip cookies can she buy?

9. Bethany has 88¢. How many peanut butter cookies can she buy?

10. Don has 72¢. How many oatmeal raisin cookies can he buy?

Bonus: How much would it cost for one dozen of each kind of cookie?

Count Your Chickens

It takes ten hens one week to lay seventy eggs. If each hen lays an equal number of eggs each day, complete the division story problems below.

1. How long will it take five hens to lay seventy eggs?

2. How long will it take one hen to lay half a dozen eggs?

3. How long will it take two hundred hens to lay two hundred eggs?

4. How many hens are needed to lay fifty eggs in two days?

5. How long will it take ten hens to lay one hundred and forty eggs?

Bonus: If half of the ten hens laid brown speckled eggs, how many hens could truthfully state that they lay white eggs?

All the Marbles

Four children collect marbles. Use the facts below to solve the division story problems.

1. Which children could evenly divide their marbles into five piles?

2. Which children could evenly divide their marbles into eleven piles?

3. Which children could evenly divide their marbles into six piles?

4. Which children could evenly divide their marbles into two piles?

5. Which children could evenly divide their marbles into three piles?

6. Which children could evenly divide their marbles into four piles?

Bonus: Which boy could evenly divide his marbles eight different ways?

Break the Bank

Write a division problem with a quotient for each problem below.

1. How many quarters in a dollar?

2. How many nickels in a half dollar?

3. How many dimes in one dollar?

4. How many pennies in one dime?

5. How many pennies in one quarter?

6. How many pennies in one nickel?

7. How many dimes in five dollars?

8. How many quarters in half a dollar?

9. How many quarters in five dollars?

10. How many pennies in ten dollars?

Bonus: Can you think of two different ways to use exactly fifty coins that equal five dollars?

Dot-to-Dot Division

Complete the dot-to-dot picture by connecting the quotients in the appropriate order. To discover the order for connecting the dots, work each division problem below.

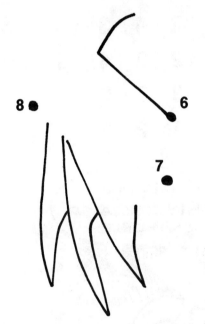

1. $144 \div 12 =$
2. $132 \div 12 =$
3. $48 \div 12 =$
4. $56 \div 7 =$
5. $99 \div 11 =$
6. $49 \div 7 =$
7. $12 \div 12 =$
8. $42 \div 7 =$
9. $55 \div 11 =$
10. $18 \div 6 =$
11. $8 \div 4 =$
12. $0 \div 1 =$
13. $28 \div 2 =$
14. $26 \div 2 =$
15. $50 \div 2 =$
16. $100 \div 10 =$

Bonus: Color your dot-to-dot division drawing.

Hide-and-Go-Seek Division

In each row below, the division problem is followed by a string of numbers. Hidden within those numbers are three division problems with the same quotient as the first problem in the row. Write each quotient and then circle the number pairs with the same quotient in each row. The first one has been completed for you.

1. 12 ÷ 3 = 4   (1   6 ÷ 4)   7   (8 ÷ 2)   (8 ÷ 7)   7
2. 25 ÷ 5 =   3   0   6   0   1   2   0   4   6
3. 18 ÷ 6 =   3   3   1   1   2   1   7   6   2
4. 42 ÷ 7 =   4   4   8   8   1   8   3   6   1
5. 64 ÷ 8 =   7   2   9   6   1   2   2   4   3
6. 21 ÷ 3 =   5   6   8   6   3   9   2   8   4
7. 81 ÷ 9 =   7   2   8   5   4   6   3   7   9
8. 48 ÷ 4 =   3   6   3   1   4   4   1   2   1

READY OR NOT HERE I COME

NINE WILL NEVER FIND ME CLEVERLY CONCEALED AS A DIVISION PROBLEM.

Bonus: Can you find and circle three division problems with the same quotient hidden in this string of numbers?

3   6   9   0   6   4   8   0

Signing Division

Use the sign language for the numbers 0-10 to complete the division problems below.

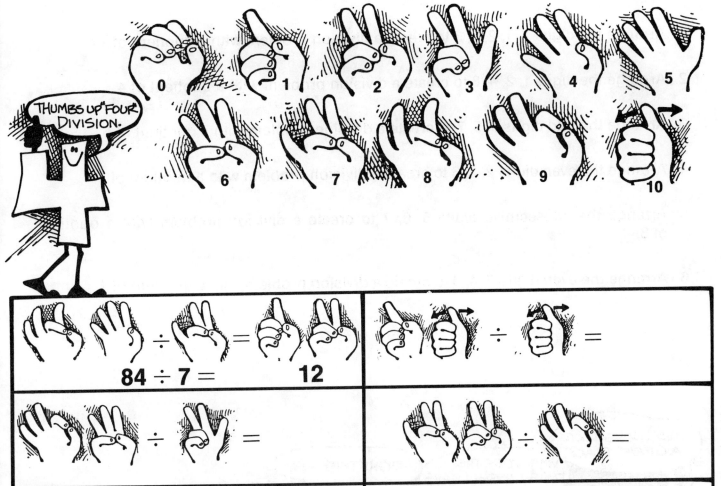

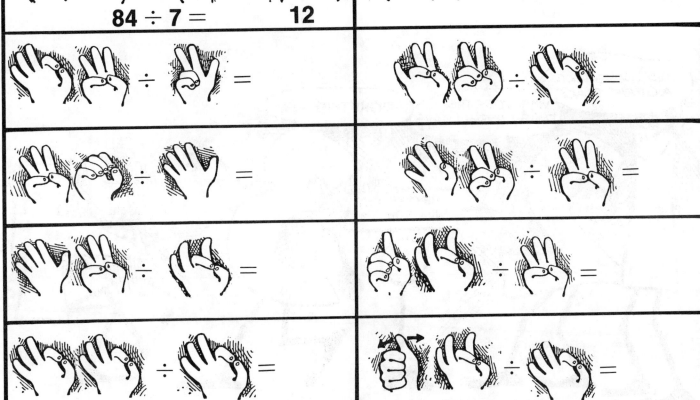

Bonus: Learn to sign your telephone number.

Making Arrangements

Follow the directions below carefully. Example: Arrange the digits 1, 1, 2, 2 to create a division problem with a quotient of 2. 22 ÷ 11 = 2

1. Arrange the digits 1, 1, 2, 2, 3 to create a division problem with a quotient of 11.

2. Arrange the digits 1, 2, 4, 8 to create a division problem with a quotient of 4.

3. Arrange the odd digits 3, 5, 7 to create a division problem with a quotient of 5.

4. Arrange the even digits 2, 4, 6 to create a division problem with a quotient of 7.

5. Arrange the consecutive digits 5, 6, 7 to create a division problem with a quotient of 8.

6. Arrange the even digits 2, 4, 8 to create a division problem with a quotient of 7.

Bonus: Arrange four digits that are the same to create a division problem that has the same quotient as the digits used in the problem.

The Magic Quilt

Color the sections of the quilt containing a number that can evenly be divided by 3 RED. Color sections of the quilt containing a number that can evenly be divided by 5 BLUE. Color sections of the quilt containing a number that can evenly be divided by 7 YELLOW. Don't be surprised if some quilt sections come up green, purple or orange!

1	2	3	4	5	6	7	8	9	10	11	12
13	14	15	16	17	18	19	20	21	22	23	24
25	26	27	28	29	30	31	32	33	34	35	36
37	38	39	40	41	42	43	44	45	46	47	48
49	50	51	52	53	54	55	56	57	58	59	60
61	62	63	64	65	66	67	68	69	70	71	72
73	74	75	67	77	78	79	80	81	82	83	84
85	86	87	88	89	90	91	92	93	94	95	96
97	98	99	100	101	102	103	104	105	106	107	108
109	110	111	112	113	114	115	116	117	118	119	120
121	122	123	124	125	126	127	128	129	130	131	132
133	134	135	136	137	138	139	140	141	142	143	144

TOGETHER WE GET ALL PURPLE!

YOU'RE SO TRANSPARENT!

Bonus: How many sections of the quilt did you color green? Purple? Orange?

Calendar Magic

1. Draw a box around any four dates on the calendar to form a square. In the example the numbers chosen were 16, 17, 23, 24.
2. Add the four numbers inside the square.
   16 + 17 + 23 + 24 = 80
3. Divide the number by 4. 80 ÷ 4 = 20
4. Subtract 4. 20 − 4 = 16
5. Compare the difference with the number at the top left-hand corner in the box that you drew in step 1.
6. Choosing four new numbers each time, repeat steps 1-5 in each of the boxes below.

MARCH

M	T	W	T	F	S	S
1	2	3	4	5	6	7
8	9	10	11	12	13	14
15	16	17	18	19	20	21
22	23	24	25	26	27	28
29	30	31				

SEPTEMBER

M	T	W	T	F	S	S
1	2	3	4	5	6	7
8	9	10	11	12	13	14
15	16	17	18	19	20	21
22	23	24	25	26	27	28
29	30					

OCTOBER

M	T	W	T	F	S	S
		1	2	3	4	5	6
7	8	9	10	11	12	13
14	15	16	17	18	19	20
21	22	23	24	25	26	27
28	29	30	31			

DECEMBER

M	T	W	T	F	S	S
			1	2	3	4
5	6	7	8	9	10	11
12	13	14	15	16	17	18
19	20	21	22	23	24	25
26	27	28	29	30	31	

FEBRUARY

M	T	W	T	F	S	S
			1	2	3	4
5	6	7	8	9	10	11
12	13	14	15	16	17	18
19	20	21	22	23	24	25
26	27	28				

I KNOW WHY THIS WORKS.

Bonus: Explain why you think this magic trick works each time.

One More Time, Please

1. Choose any three-digit number. Example: 832
2. Add 25 to the three-digit number. Example: 832 + 25 = 857
3. Multiply the sum by 2. Example: 857 × 2 = 1714
4. Subtract 4 from the product. Example: 1714 − 4 = 1710
5. Divide the answer by 2. Example: 1710 ÷ 2 = 855
6. Subtract the three-digit number you picked in step 1 from the difference. Example: 855 − 832 = 23
7. Choosing different three-digit numbers, repeat steps 1-6 in each box below.

Bonus: What pattern did you discover? Make up your own seven-step magic number trick.

Fingers on One Hand

1. Choose any three-digit number. Example: 742
2. Add the next higher number. Example: 742 + 743 = 1485
3. Add 9 to the total. Example: 1485 + 9 = 1494
4. Divide the total by 2. Example: 1494 ÷ 2 = 747
5. Subtract the original number from the answer. Example: 747 − 742 = 5
6. Choosing a different three-digit number each time, repeat steps 1-5 in each of the boxes below.

Bonus: What pattern did you discover? Explain the reason for the pattern.

Pick a Number, Any Number

1. Choose any number. Example: 3
2. Add 30 to the number. Example: 3 + 30 = 33
3. Multiply the sum by 2. Example: 33 × 2 = 66
4. Subtract 4 from the product. Example: 66 − 4 = 62
5. Divide the answer by 2. Example: 62 ÷ 2 = 31
6. Subtract 28 from the answer. Example: 31 − 28 = 3
7. Choosing a different number each time, repeat steps 1-6 in each box below.

Bonus: Show this magic trick to three friends.

$1 \div 1 =$	$22 \div 2 =$	$12 \div 3 =$
$33 \div 3 =$	$16 \div 4 =$	$44 \div 4 =$
$18 \div 6 =$	$36 \div 6 =$	$81 \div 9 =$
$45 \div 9 =$	$66 \div 11 =$	$90 \div 9 =$
$132 \div 12 =$	$96 \div 12 =$	$108 \div 12 =$
$54 \div 6 =$	$14 \div 7 =$	$16 \div 8 =$
$3 \div 3 =$	$8 \div 2 =$	$11 \div 1 =$
$21 \div 3 =$	$36 \div 3 =$	$4 \div 4 =$
$50 \div 5 =$	$24 \div 6 =$	$27 \div 9 =$
$54 \div 9 =$	$90 \div 10 =$	$55 \div 11 =$

120 ÷ 12 =	84 ÷ 12 =	36 ÷ 12 =
60 ÷ 6 =	35 ÷ 7 =	21 ÷ 7 =
20 ÷ 2 =	24 ÷ 3 =	6 ÷ 2 =
9 ÷ 1 =	18 ÷ 2 =	30 ÷ 5 =
6 ÷ 6 =	36 ÷ 4 =	55 ÷ 5 =
28 ÷ 4 =	32 ÷ 4 =	60 ÷ 5 =
110 ÷ 10 =	121 ÷ 11 =	88 ÷ 11 =
100 ÷ 10 =	80 ÷ 8 =	49 ÷ 7 =
63 ÷ 9 =	64 ÷ 8 =	15 ÷ 3 =
40 ÷ 4 =	9 ÷ 3 =	25 ÷ 5 =

70 ÷ 7 =	99 ÷ 11 =	132 ÷ 11 =
12 ÷ 2 =	24 ÷ 2 =	27 ÷ 3 =
30 ÷ 3 =	48 ÷ 4 =	20 ÷ 4 =
45 ÷ 5 =	30 ÷ 6 =	42 ÷ 6 =
10 ÷ 2 =	4 ÷ 2 =	5 ÷ 1 =
72 ÷ 6 =	56 ÷ 7 =	24 ÷ 8 =
10 ÷ 10 =	40 ÷ 8 =	28 ÷ 7 =
20 ÷ 10 =	120 ÷ 10 =	9 ÷ 9 =
18 ÷ 9 =	144 ÷ 12 =	12 ÷ 12 =
48 ÷ 6 =	42 ÷ 7 =	77 ÷ 7 =

Graphing Your Progress

Number Correct													
90													
88													
86													
84													
82													
80													
78													
76													
74													
72													
70													
68													
66													
64													
62													
60													
58													
56													
54													
52													
50													
48													
46													
44													
42													
40													
38													
36													
34													
32													
30													
28													
26													
24													
22													
20													
18													
16													
14													
12													
10													
8													
6													
4													
2													
Test #	1	2	3	4	5	6	7	8	9	10	11	12	13

Division Bingo

Object: To practice the division facts 1 ÷ 1 through 144 ÷ 12. This game is for two or more players and works well with a large group.

Rules: To play this game you must begin by filling in the Bingo card with appropriate numbers. Under the letter B place five numbers, one in each box, from the following: 1, 2, 3, 4, 5, 6, 7, 8, 9, 10, 11, 12.

Under the letter I place five numbers, one in each box, from the following: 14, 15, 16, 18, 20, 21, 22, 24, 25, 27, 28, 30.

Under the letter N place four numbers, one in each box, from the following: 32, 33, 35, 36, 40, 42, 44, 45, 48, 49, 50, 54.

Under the letter G place five numbers, one in each box, from the following: 55, 56, 60, 63, 64, 66, 70, 72, 77, 80, 81, 84, 88.

Under the letter O place five numbers, one in each box, from the following: 90, 96, 99, 100, 108, 110, 120, 121, 132, 144. Now your game card is ready. Caller rolls a pair of dice and calls the number indicated by the dice. If the player has a number on his/her card that can be divided evenly by the number indicated on the dice, he/she writes the division problem in the appropriate space. Example: If a player has placed the number 144 under the letter O on his/her card and the caller rolls two 6's and calls out the number 12, the player can write 144 ÷ 12 = 12 in the appropriate box on his/her game card. The first player to have a division problem written in all the spaces on his/her card shouts, "Bingo!" and is declared the winner. To shorten the game, players must have a division problem in five spaces in a straight line, horizontally, vertically or diagonally.

Division Bingo Card

"Well, it's about time letters come "B"-4 numbers for a change."

"But I'm afraid all the answers are numbers."

"I'm going to N-joy myself N-yway."

"G. U.R.A.Q.T."

"O my, letters are very punny."

"That's the point!"

BINGO

Top Eighty!

Object: To practice the division facts 1 ÷ 1 through 36 ÷ 6

Rules: Roll a pair of dice twenty times. Each time you roll, think of the number that can be divided by one number indicated on one die and has a quotient equal to the number indicated on the other die. For example: If you roll a 6 and a 5, you would think of a number that 5 divides into and has a quotient of 6—30. Using the larger of the two numbers indicated on the dice as the quotient, record each problem on the score card below. After twenty rolls, total the quotients. If your score is 80 or more, you win. If your score is 79 or less, you lose.

#	Division Problem	Quotient
1.		
2.		
3.		
4.		
5.		
6.		
7.		
8.		
9.		
10.		
11.		
12.		
13.		
14.		
15.		
16.		
17.		
18.		
19.		
20.		
Game Total		

A SIX AND A FIVE—YES. LET'S SEE... 30÷5 IS THE PROBLEM AND 6 IS THE QUOTIENT. AT THIS RATE I'M SURE TO WIN—I HOPE YOU ARE AS LUCKY AS ME!

THIS COULD BE A LONG DAY.

Quotients Maze

Object: To practice division facts 1 ÷ 1 through 144 ÷ 12

Rules: To play this game you will need division flash cards (see pages 62, 63 and 64), a die, a marker for each player and the gameboard below. Players take turns turning over the top flash card and giving the correct quotient within five seconds. Another player can slowly count 1001, 1002, 1003, 1004, 1005, at which time the player must give his/her answer. If player gives the correct quotient, he/she rolls the die and advances marker the appropriate number of spaces on the board. The first player to reach the end of the maze is declared the winner. To avoid competition, game can continue until every player has reached the end of the maze.

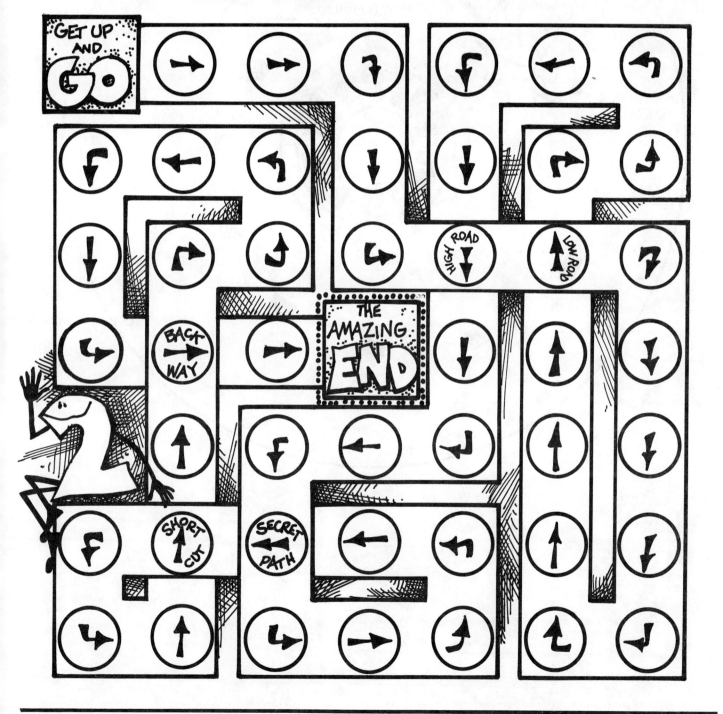

Reproducible Facts Wheel
Division

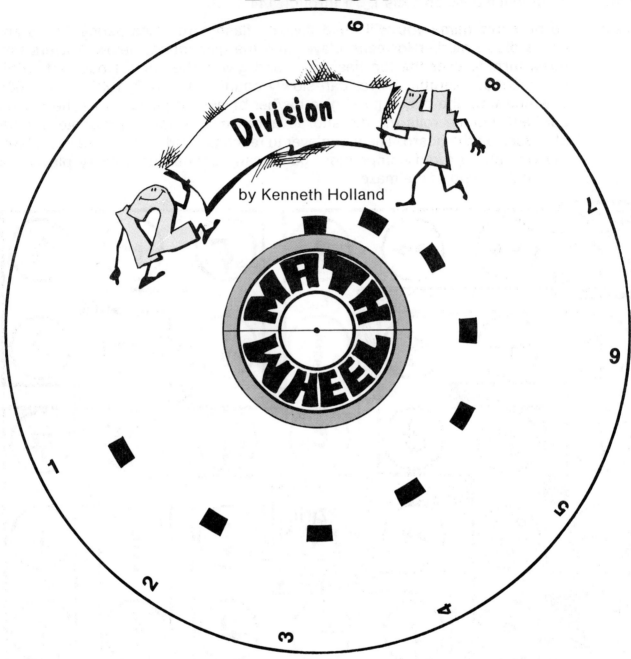

by Kenneth Holland

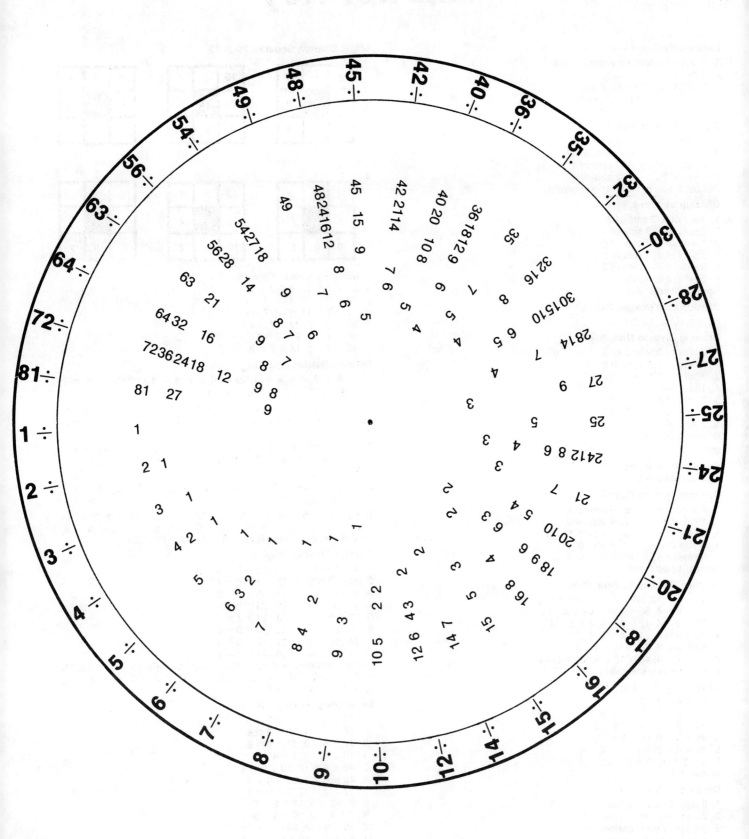

Answer Key

Dividing by One, Page 2
1. 2 pockets with 1 marble in each
2. 3 days
3. 5 bags with 1 apple in each
4. 4 boxes with 1 ball in each
Bonus: 1 set of 10 things

Dividing by Two, Page 3
1. 5 sets of 2
2. 4 sets of 2
3. 6 sets of 2
4. 1 set of 2
5. 8 sets of 1
6. 2 sets of 5
7. 5 sets of 1
8. 2 sets of 4
Bonus: 2 sets of 10 and 10 sets of 2

Dividing by Three, Page 4
1. 1 set of 3, 2 sets of 3
2. 3 sets of 3, 3 sets of 4
3. 5 sets of 2, 4 sets of 3
4. 2 sets of 5, 3 sets of 2
5. 6 sets of 3, 7 sets of 3
6. 3 sets of 6, 0 sets
Bonus: 20 sets of 5

Division Line Design, Page 5
Bonus: 6, 12, 18, 24, etc.

Drawing Division Dice, Page 6
1. 2 sets of 6, 2 sets of 5
2. 2 sets of 1, 2 sets of 3
3. 2 sets of 2, 2 sets of 4
4. 12 sets of 1
5. 3 sets of 2
6. 6 sets of 2
7. 3 sets of 4
Bonus: 6 sets of 5

Matchup, Page 7
Secret Message: study
Bonus: memorize

Dividing by Four, Page 8
1. 2 sets of 4
2. 1 set of 4
3. 3 sets of 4
4. 4 sets of 4
5. 6 sets of 4
6. 5 sets of 4
7. 4 sets of 5
8. 4 sets of 6
9. 4 sets of 3
10. 4 sets of 2
Bonus: 12 sets of 4

Practice Dividing by Four, Page 9
1. 48 ÷ 4 = 12, 28 ÷ 4 = 7, 32 ÷ 4 = 8
2. 24 ÷ 6 = 4, 36 ÷ 4 = 9, 44 ÷ 4 = 11
3. 16 ÷ 4 = 4, 12 ÷ 3 = 4, 4 ÷ 1 = 4
4. 36 ÷ 9 = 4, 24 ÷ 4 = 6, 8 ÷ 2 = 4
5. 12 ÷ 4 = 3, 8 ÷ 4 = 2, 4 ÷ 4 = 1
6. 28 ÷ 7 = 4, 48 ÷ 12 = 4, 32 ÷ 8 = 4
Bonus: Answers will vary.

Dividing by Five, Page 10
1. 25 ÷ 5 = 5, 35 ÷ 5 = 7
2. 30 ÷ 5 = 6, 10 ÷ 5 = 2
3. 60 ÷ 5 = 12, 40 ÷ 5 = 8
4. 45 ÷ 5 = 9, 50 ÷ 5 = 10
5. 15 ÷ 5 = 3, 20 ÷ 5 = 4
6. 5 ÷ 5 = 1, 55 ÷ 5 = 11
Bonus: 20

Dividing by Six, Page 12
1. 6 eggs in one carton
2. 2 eggs in each carton
3. 3 eggs in each carton
4. 1 egg in each carton
Bonus: Ten dozen divided by half a dozen is the same as 120 ÷ 6 = 20. Five dozen divided by ten is the same as 60 ÷ 10 = 6. So, ten dozen divided by half a dozen is more than five dozen divided by ten.

Magic Division Squares, Page 13

1.
2.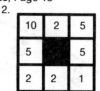
3.
4.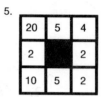
5.
6.

Dividing by Seven, Page 14
1. 21 ÷ 7 = 3, 70 ÷ 7 = 10
2. 84 ÷ 7 = 12, 49 ÷ 7 = 7
3. 77 ÷ 7 = 11, 63 ÷ 7 = 9
4. 35 ÷ 7 = 5, 28 ÷ 7 = 4
5. 56 ÷ 7 = 8, 7 ÷ 7 = 1
6. 42 ÷ 7 = 6, 14 ÷ 7 = 2

Division Daisies, Page 15
32 ÷ 4 = 8, 4 ÷ 4 = 1, 28 ÷ 4 = 7, 8 ÷ 4 = 2, 36 ÷ 4 = 9, 24 ÷ 4 = 6, 12 ÷ 4 = 3, 20 ÷ 4 = 5, 16 ÷ 4 = 4
10 ÷ 2 = 5, 6 ÷ 2 = 3, 4 ÷ 2 = 2, 8 ÷ 2 = 4, 20 ÷ 2 = 10, 18 ÷ 2 = 9, 0 ÷ 2 = 0, 2 ÷ 2 = 1, 12 ÷ 2 = 6, 14 ÷ 2 = 7, 16 ÷ 2 = 8
30 ÷ 3 = 10, 9 ÷ 3 = 3, 6 ÷ 3 = 2, 18 ÷ 3 = 6, 12 ÷ 3 = 4, 24 ÷ 3 = 8, 15 ÷ 3 = 5, 3 ÷ 3 = 1, 21 ÷ 3 = 7, 27 ÷ 3 = 9
20 ÷ 5 = 4, 30 ÷ 5 = 6, 35 ÷ 5 = 7, 25 ÷ 5 = 5, 15 ÷ 5 = 3, 10 ÷ 5 = 2, 5 ÷ 5 = 1
Bonus: 0 ÷ 6 = 0, 6 ÷ 6 = 1, 12 ÷ 6 = 2, 18 ÷ 6 = 3, 24 ÷ 6 = 4, 30 ÷ 6 = 5, 36 ÷ 6 = 6, 42 ÷ 6 = 7, 48 ÷ 6 = 8, 54 ÷ 6 = 9, 60 ÷ 6 = 10, 66 ÷ 6 = 11, 72 ÷ 6 = 12

Charter Division, Page 17
1. 4 ÷ 1 = 4
2. 10 ÷ 2 = 5
3. 8 ÷ 4 = 2
4. 9 ÷ 3 = 3
5. 10 ÷ 5 = 2
6. 8 ÷ 2 = 4
7. 15 ÷ 3 = 5
8. 16 ÷ 2 = 8
9. 20 ÷ 5 = 4
10. 24 ÷ 6 = 4
11. 20 ÷ 4 = 5
12. 24 ÷ 8 = 3
13. 21 ÷ 7 = 3
14. 10 ÷ 5 = 2
15. 12 ÷ 4 = 3
16. 21 ÷ 3 = 7
Bonus: Answers will vary. Some may include 20 ÷ 2 = 10, 16 ÷ 8 = 2, 16 ÷ 4 = 4, 20 ÷ 10 = 2, 12 ÷ 6 = 2, 12 ÷ 2 = 6, 12 ÷ 3 = 4, 20 ÷ 4 = 5.

Letter Division, Page 18
1. 48 ÷ 6 = 8, 35 ÷ 7 = 5
2. 28 ÷ 7 = 4, 32 ÷ 8 = 4
3. 72 ÷ 8 = 9, 54 ÷ 6 = 9
4. 16 ÷ 8 = 2, 4 ÷ 4 = 1
5. 40 ÷ 4 = 10, 20 ÷ 5 = 4
6. 42 ÷ 6 = 7, 30 ÷ 6 = 5
7. 12 ÷ 1 = 12, 2 ÷ 2 = 1
8. 9 ÷ 3 = 3, 18 ÷ 3 = 6
9. 8 ÷ 4 = 2, 36 ÷ 4 = 9
10. 45 ÷ 5 = 9, 50 ÷ 5 = 10
11. 60 ÷ 6 = 10, 24 ÷ 6 = 4
12. 4 ÷ 2 = 2, 21 ÷ 3 = 7

Coloring Division, Page 19
Bonus: 1. 35 can evenly be divided by 5 and 7. 2. 30 can evenly be divided by 5 and 6. 3. 42 can evenly be divided by 6 and 7.

Dividing by Nine, Page 20
1. 108 ÷ 9 = 12, 99 ÷ 9 = 11
2. 54 ÷ 9 = 6, 63 ÷ 9 = 7
3. 36 ÷ 9 = 4, 45 ÷ 9 = 5
4. 9 ÷ 9 = 1, 18 ÷ 9 = 2
5. 27 ÷ 9 = 3, 72 ÷ 9 = 8
Bonus: Totals always equal 9.

Division Families, Page 21
1,1; 2,2; 3,3; 4,4; 5,5; 6,6; 7,7; 8,8; 9,9; 10,10; 11,11; 12,12
2,1; 4,2; 6,3; 8,4; 10,5; 12,6; 14,7; 16,8; 18,9; 20,10; 22,11; 24,12
3,1; 6,2; 9,3; 12,4; 15,5; 18,6; 21,7; 24,8; 27,9; 30,10; 33,11; 36,12
4,1; 8,2; 12,3; 16,4; 20,5; 24,6; 28,7; 32,8; 36,9; 40,10; 44,11; 48,12
5,1; 10,2; 15,3; 20,4; 25,5; 30,6; 35,7; 40,8; 45,9; 50,10; 55,11; 60,12
6,1; 12,2; 18,3; 24,4; 30,5; 36,6; 42,7; 48,8; 54,9; 60,10; 66,11; 72,12
7,1; 14,2; 21,3; 28,4; 35,5; 42,6; 49,7; 56,8; 63,9; 70,10; 77,11; 84,12
8,1; 16,2; 24,3; 32,4; 40,5; 48,6; 56,7; 64,8; 72,9; 80,10; 88,11; 96,12
9,1; 18,2; 27,3; 36,4; 45,5; 54,6; 63,7; 72,8; 81,9; 90,10; 99,11; 108,12
Bonus: 15:1, 2, 3, 12, 13, 21, 23, 31, 32, 123, 132, 213, 231, 312, 321

About Face!, Page 22
1. 72 ÷ 9 = 8, 72 ÷ 8 = 9
2. 54 ÷ 9 = 6, 54 ÷ 6 = 9
3. 63 ÷ 7 = 9, 63 ÷ 9 = 7
4. 45 ÷ 5 = 9, 45 ÷ 9 = 5
5. 18 ÷ 9 = 2, 18 ÷ 2 = 9
6. 27 ÷ 3 = 9, 27 ÷ 9 = 3
7. 90 ÷ 10 = 9, 90 ÷ 9 = 10
8. 99 ÷ 9 = 11, 99 ÷ 11 = 9
9. 36 ÷ 9 = 4, 36 ÷ 4 = 9
10. 108 ÷ 12 = 9, 108 ÷ 9 = 12
Bonus: 81

Division Pairs, Page 23
1. eight of any of the following number pairs: 6,1; 12,2; 18,3; 24,4; 30,5; 36,6; 42,7; 48,8; 54,9; 60,10; 66,11; 72,12
2. eight of any of the following number pairs: 7,1; 14,2; 21,3; 28,4; 35,5; 42,6; 49,7; 56,8; 63,9; 70,10; 77,11; 84,12
3. eight of any of the following number pairs: 8,1; 16,2; 24,3; 32,4; 40,5; 48,6; 56,7; 64,8; 72,9; 80,10; 88,11; 96,12
4. eight of any of the following number pairs: 9,1; 18,2; 27,3; 36,4; 45,5; 54,6; 63,7; 72,8; 81,9; 90,10; 99,11; 108,12
Bonus: eight of any of the following number pairs: 12,1; 24,2; 36,3; 48,4; 60,5; 72,6; 84,7; 96,8; 108,9; 120,10; 132,11; 144,12

Dividing by Tens, Page 24
60 ÷ 10 = 6, 6 sets of 10 dots
50 ÷ 10 = 5, 5 sets of 10 dots
70 ÷ 7 = 10, 10 sets of 7 dots
90 ÷ 10 = 9, 9 sets of 10 dots
40 ÷ 10 = 4, 4 sets of 10 dots
20 ÷ 2 = 10, 10 sets of 2 dots
80 ÷ 8 = 10, 10 sets of 8 dots
100 ÷ 10 = 10, 10 sets of 10 dots
10 ÷ 1 = 10, 10 sets of 1 dot
Bonus: 10 sets of 12 dots

Dividing by Eleven, Page 25
1. 9, 10
2. 1, 3
3. 11, 11
4. 11, 6
5. 7, 11
6. 11, 11
7. 11, 11
8. 11, 12
Bonus: double numbers

Perfect Pairs, Page 26

Division Twisters, Page 27
1. 66 ÷ 6 = 11
2. 55 ÷ 5 = 11
3. 88 ÷ 8 = 11
4. 44 ÷ 4 = 11
5. 99 ÷ 9 = 11
Bonus: 3 miles

Dividing by Twelve, Page 28
1. 24 ÷ 2 = 12, 12 sets of 2
2. 96 ÷ 8 = 12, 12 sets of 8
3. 36 ÷ 3 = 12, 12 sets of 3
4. 84 ÷ 7 = 12, 12 sets of 7
5. 60 ÷ 5 = 12, 12 sets of 5
6. 48 ÷ 4 = 12, 12 sets of 4
7. 132 ÷ 11 = 12, 12 sets of 11
8. 12 ÷ 1 = 12, 12 sets of 1
Bonus: 6 eggs in 6 sections of the carton

Division Dart Board, Page 29

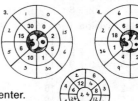

Bonus: 48 may also be used in the center.

Paths of Division, Page 30
1. 132 ÷ 11 = 12
2. 121 ÷ 11 = 11
3. 99 ÷ 11 = 9
4. 108 ÷ 9 = 12
5. 96 ÷ 12 = 8
6. 144 ÷ 12 = 12

Ring Three, Page 31

144	12	12	132	12	11
1	1	1	2	1	2
56	63	7	9	84	8
7	70	7	10	7	8
8	77	7	11	12	1
120	12	10	108	12	9

120	12	10	108	12	9
96	84	12	7	72	60
12	24	12	2	12	12
8	12	12	1	6	5
48	12	4	36	12	3

5	1	5	33	55	60
24	3	8	3	5	12
27	30	36	11	11	5
3	3	3	44	11	4
9	10	12	48	4	2

3	2	2	1	4	14
1	12	2	6	4	2
3	8	2	2	1	7
4	1	4	8	2	4
10	2	5	12	2	6

Bonus:

0	6	0	3	0	3
3	6	2	3	1	3
0	1	1	1	1	1
2	1	2	3	1	3
2	0	0	0	0	0
1	0	0	6	0	0

Division Towers, Page 32

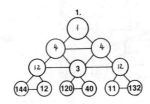
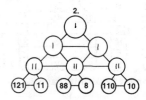
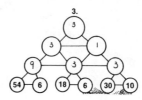
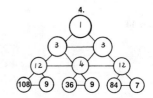

The Division Connection, Page 33
Secret Message: One cuts, the other chooses his piece.
Bonus: coconut cream

Heart-Shaped Division, Page 34
1. $77 \div 7 = 11$, $11 \div 1 = 1$
2. $9 \div 1 = 9$, $63 \div 7 = 9$
3. $48 \div 4 = 12$, $88 \div 11 = 8$, $25 \div 5 = 5$
4. $44 \div 4 = 11$, $72 \div 6 = 12$, $96 \div 8 = 12$
5. $84 \div 7 = 12$, $7 \div 7 = 1$, $88 \div 11 = 8$
6. $99 \div 11 = 9$, $88 \div 8 = 11$, $72 \div 9 = 8$

Coded Division, Page 35
1. $120 \div 12 = 10$, $48 \div 12 = 4$
2. $15 \div 5 = 3$, $20 \div 5 = 4$
3. $6 \div 3 = 2$, $36 \div 12 = 3$
4. $12 \div 6 = 2$, $12 \div 12 = 1$
5. $24 \div 4 = 6$, $24 \div 12 = 2$
6. $30 \div 6 = 5$, $25 \div 5 = 5$
7. $64 \div 8 = 8$, $18 \div 6 = 3$
8. $24 \div 8 = 3$, $72 \div 8 = 9$

Computer Division, Page 36
1. $16 \div 2 = 8$, $8 + 9 = 17$
2. $10 \div 2 = 5$, $5 + 9 = 14$
3. $15 \div 5 = 3$, $3 + 9 = 12$
4. $35 \div 5 = 7$, $7 + 9 = 16$
5. $14 \div 2 = 7$, $7 + 9 = 16$
6. $45 \div 5 = 9$, $9 + 9 = 18$
7. $22 \div 2 = 11$, $11 + 9 = 20$
8. $24 \div 2 = 12$, $12 + 9 = 21$
9. $5 \div 5 = 1$, $1 + 9 = 10$
10. $55 \div 5 = 11$, $11 + 9 = 20$
11. $20 \div 2 = 10$, $10 + 9 = 19$
12. $8 \div 2 = 4$, $4 + 9 = 13$
Bonus: 25 ($25 \div 5 = 5$, $5 + 9 = 14$)

Double-Cross Division, Page 37
1. $144 \div 12 = 12$, $36 \div 12 = 3$
2. $54 \div 6 = 9$, $72 \div 12 = 6$
3. $24 \div 12 = 2$, $30 \div 10 = 3$
4. $55 \div 11 = 5$, $132 \div 12 = 11$
5. $24 \div 12 = 2$, $40 \div 5 = 8$
6. $72 \div 6 = 12$, $36 \div 12 = 3$
7. $7 \div 7 = 1$, $66 \div 11 = 6$
8. $32 \div 4 = 8$, $14 \div 2 = 7$
Bonus:
$>\vee\vee \div >\vee = >\vee$,
$>\wedge> \div >> = >>$

Search and Circle, Page 38

Common Quotients, Page 39
1. 3
2. 7
3. 10
4. 4
5. 12
6. 5
7. 11
8. 9
Bonus: 43

What's My Sign?, Page 40
1. $48 \div 6 = 8$, $60 \div 12 = 5$
2. $84 \div 12 = 7$, $54 \div 6 = 9$
3. $60 \div 6 = 10$, $96 \div 12 = 8$
4. $63 \div 7 = 9$, $70 \div 7 = 10$
5. $120 \div 12 = 10$, $28 \div 7 = 4$
6. $35 \div 7 = 5$, $132 \div 12 = 11$
7. $49 \div 7 = 7$, $72 \div 12 = 6$
8. $64 \div 8 = 8$, $36 \div 12 = 3$
9. $48 \div 12 = 4$, $42 \div 7 = 6$
10. $72 \div 8 = 9$, $132 \div 11 = 12$
11. $24 \div 12 = 2$, $120 \div 12 = 10$
12. $56 \div 8 = 7$, $56 \div 7 = 8$
13. $144 \div 12 = 12$, $48 \div 8 = 6$
14. $24 \div 6 = 4$, $18 \div 6 = 3$
15. $42 \div 6 = 7$, $120 \div 10 = 12$
16. $121 \div 11 = 11$, $110 \div 11 = 10$

Arrow Division, Page 41
1. $20 \div 2 = 10$, $20 \div 4 = 5$, $20 \div 10 = 2$
2. $20 \div 5 = 4$, $10 \div 5 = 2$, $10 \div 2 = 5$
3. $12 \div 2 = 6$, $12 \div 4 = 3$, $18 \div 2 = 9$
4. $8 \div 2 = 4$, $12 \div 3 = 4$, $4 \div 2 = 2$
5. $18 \div 3 = 6$, $18 \div 2 = 9$, $10 + 20 \div 5 = 6$

Connect Them All!, Page 42

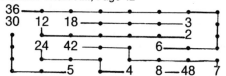

Hide and Seek Problems, Page 44
1. $99 \div 9 = 11$ and $99 \div 11 = 9$,
$108 \div 9 = 12$ and $108 \div 12 = 9$,
$120 \div 10 = 12$ and $120 \div 12 = 10$,
$44 \div 11 = 4$ and $44 \div 4 = 11$
2. $48 \div 4 = 12$ and $48 \div 12 = 4$,
$55 \div 11 = 5$ and $55 \div 5 = 11$,
$24 \div 6 = 4$ and $24 \div 4 = 6$,
$30 \div 5 = 6$ and $30 \div 6 = 5$
3. $56 \div 8 = 7$ and $56 \div 7 = 8$,
$48 \div 8 = 6$ and $48 \div 6 = 8$,
$32 \div 8 = 4$ and $32 \div 4 = 8$ or $24 \div 8 = 3$ and $24 \div 3 = 8$,
$84 \div 12 = 7$ and $84 \div 7 = 12$
Bonus: $100 \div 0 = 0$

Color It Division, Page 45
Bonus: 1. 96, 2. 108

Checking Division, Page 46
10: 1,10; 2,5
12: 1,12; 2,6; 3,4
16: 2,8; 4,4
18: 2,9; 3,6
20: 2,10; 4,5
24: 2,12; 3,8; 4,6
30: 5,6; 3,10
Bonus: Answers may vary. One possible answer is 36: 3,12; 4,9; 6,6.

Just Two Little Lines, Page 47

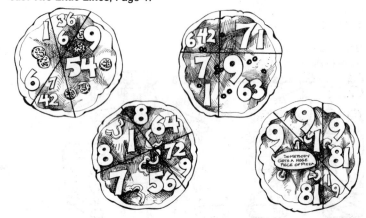

Just Three Lines!, Page 48

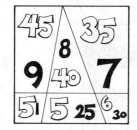

The Bake Sale, Page 49
1. $.64 ÷ .08 = 8
2. $1.00 ÷ .10 = 10
3. $1.44 ÷ .12 = 12
4. $.72 ÷ .09 = 8
5. $1.44 ÷ .08 = 18
6. $1.44 ÷ .12 = 12
7. $1.32 ÷ .12 = 11
8. $1.50 ÷ .10 = 15
9. $.88 ÷ .08 = 11
10. $.72 ÷ .09 = 8
Bonus: $4.68

Count Your Chickens, Page 50
1. 2 weeks
2. 6 days
3. 1 day
4. 25
5. 2 weeks
Bonus: None, hens cannot speak.

All the Marbles, Page 51
1. Bill, James and John
2. Max and James
3. John and Max
4. All four boys
5. John and Max
6. John and Max
Bonus: John: 2, 3, 4, 5, 6, 8, 10 and 12

Break the Bank, Page 52
1. 100 ÷ 25 = 4
2. 50 ÷ 5 = 10
3. 100 ÷ 10 = 10
4. 10 ÷ 1 = 10
5. 25 ÷ 1 = 25
6. 5 ÷ 1 = 5
7. 500 ÷ 10 = 50
8. 50 ÷ 25 = 2
9. 500 ÷ 25 = 20
10. 1000 ÷ 1 = 1000
Bonus: 50 dimes; 25 nickels, 3 quarters, 2 half dollars and 20 dimes

Dot-to-Dot Division, Page 53
1. 144 ÷ 12 = 12
2. 132 ÷ 12 = 11
3. 48 ÷ 12 = 4
4. 56 ÷ 7 = 8
5. 99 ÷ 11 = 9
6. 49 ÷ 7 = 7
7. 12 ÷ 12 = 1
8. 42 ÷ 7 = 6
9. 55 ÷ 11 = 5
10. 18 ÷ 6 = 3
11. 8 ÷ 4 = 2
12. 0 ÷ 1 = 0
13. 28 ÷ 2 = 14
14. 26 ÷ 2 = 13
15. 50 ÷ 2 = 25
16. 100 ÷ 10 = 10

Hide-and-Go-Seek Division, Page 54
1. 12 ÷ 3 = 4
2. 25 ÷ 5 = 5
3. 18 ÷ 6 = 3
4. 42 ÷ 7 = 6
5. 64 ÷ 8 = 8
6. 21 ÷ 3 = 7
7. 81 ÷ 9 = 9
8. 48 ÷ 4 = 12

Bonus: 369 ÷ 0 = 0, 69 ÷ 0 = 0, 9 ÷ 0 = 0, 648 ÷ 0 = 0, 48 ÷ 0 = 0, 8 ÷ 0 = 0

Signing Division, Page 55
1. 84 ÷ 7 = 12, 110 ÷ 10 = 11
2. 96 ÷ 3 = 32, 72 ÷ 9 = 8
3. 60 ÷ 5 = 12, 42 ÷ 6 = 7
4. 56 ÷ 8 = 7, 18 ÷ 6 = 3
5. 99 ÷ 9 = 11, 108 ÷ 9 = 12

Making Arrangements, Page 56
1. 132 ÷ 12 = 11
2. 48 ÷ 12 = 4
3. 35 ÷ 7 = 5
4. 42 ÷ 6 = 7
5. 56 ÷ 7 = 8
6. 28 ÷ 4 = 7
Bonus: 11 ÷ 11 = 1

The Magic Quilt, Page 57

red: 3, 6, 9, 12, 15, 18, 21, 24, 27, 30, 33, 36, 39, 42, 45, 48, 51, 54, 57, 60, 63, 66, 69, 72, 75, 78, 81, 84, 87, 90, 93, 96, 99, 102, 105, 108, 111, 114, 117, 120, 123, 126, 129, 132, 135, 138, 141, 144

blue: 5, 10, 15, 20, 25, 30, 35, 40, 45, 50, 55, 60, 65, 70, 75, 80, 85, 90, 95, 100, 105, 110, 115, 120, 125, 130, 135, 140

yellow: 7, 14, 21, 28, 35, 42, 49, 56, 63, 70, 77, 84, 91, 98, 105, 112, 119, 126, 133, 140

Bonus: green: 35, 70, 105, 140; purple: 15, 30, 45, 60, 75, 90, 105, 120, 135; orange: 21, 42, 63, 84, 105, 126

One More Time, Please, Page 59
Bonus: The answer is always 23.

Fingers on One Hand, Page 60
Bonus: The answer is always 5 because in step 2 you add the original number plus one more. In step 3 you add 9, so you now have twice the original number plus 10. In step 4 you divide by 2 which gives you the original number plus 5. In the last step you subtract the original number always leaving you with the number 5.

Pick a Number, Any Number, Page 61
Bonus: The answer is always your original number.

Pre/Post Test, Page 62

1 ÷ 1 = 1	22 ÷ 2 = 11	12 ÷ 3 = 4
33 ÷ 3 = 11	16 ÷ 4 = 4	44 ÷ 4 = 11
18 ÷ 6 = 3	36 ÷ 6 = 6	81 ÷ 9 = 9
45 ÷ 9 = 5	66 ÷ 11 = 6	90 ÷ 9 = 10
132 ÷ 12 = 11	96 ÷ 12 = 8	108 ÷ 12 = 9
54 ÷ 6 = 9	14 ÷ 7 = 2	16 ÷ 8 = 2
3 ÷ 3 = 1	8 ÷ 2 = 4	11 ÷ 1 = 11
21 ÷ 3 = 7	36 ÷ 3 = 12	4 ÷ 4 = 1
50 ÷ 5 = 10	24 ÷ 6 = 4	27 ÷ 9 = 3
54 ÷ 9 = 6	90 ÷ 10 = 9	55 ÷ 11 = 5

Pre/Post Test, Page 63

120 ÷ 12 = 10	84 ÷ 12 = 7	36 ÷ 12 = 3
60 ÷ 6 = 10	35 ÷ 7 = 5	21 ÷ 7 = 3
20 ÷ 2 = 10	24 ÷ 3 = 8	6 ÷ 2 = 3
9 ÷ 1 = 9	18 ÷ 2 = 9	30 ÷ 5 = 6
6 ÷ 6 = 1	36 ÷ 4 = 9	55 ÷ 5 = 11
28 ÷ 4 = 7	32 ÷ 4 = 8	60 ÷ 5 = 12
110 ÷ 10 = 11	121 ÷ 11 = 11	88 ÷ 11 = 8
100 ÷ 10 = 10	80 ÷ 8 = 10	49 ÷ 7 = 7
63 ÷ 9 = 7	64 ÷ 8 = 8	15 ÷ 3 = 5
40 ÷ 4 = 10	9 ÷ 3 = 3	25 ÷ 5 = 5

Pre/Post Test, Page 64

70 ÷ 7 = 10	99 ÷ 11 = 9	132 ÷ 11 = 12
12 ÷ 2 = 6	24 ÷ 2 = 12	27 ÷ 3 = 9
30 ÷ 3 = 10	48 ÷ 4 = 12	20 ÷ 4 = 5
45 ÷ 5 = 9	30 ÷ 6 = 5	42 ÷ 6 = 7
10 ÷ 2 = 5	4 ÷ 2 = 2	5 ÷ 1 = 5
72 ÷ 6 = 12	56 ÷ 7 = 8	24 ÷ 3 = 8
10 ÷ 10 = 1	40 ÷ 8 = 5	28 ÷ 7 = 4
20 ÷ 10 = 2	120 ÷ 10 = 12	9 ÷ 9 = 1
18 ÷ 9 = 2	144 ÷ 12 = 12	12 ÷ 12 = 1
48 ÷ 6 = 8	42 ÷ 7 = 6	77 ÷ 7 = 11

Congratulations! you know 100% of the DIVISION FACTS

to: _____
from: _____

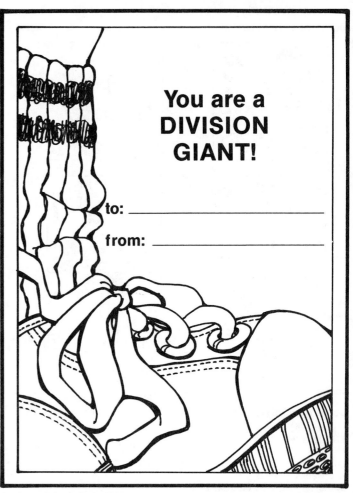

You are a DIVISION GIANT!

to: _____
from: _____

to: _____ for completing _____
Bonus activities.

from: _____

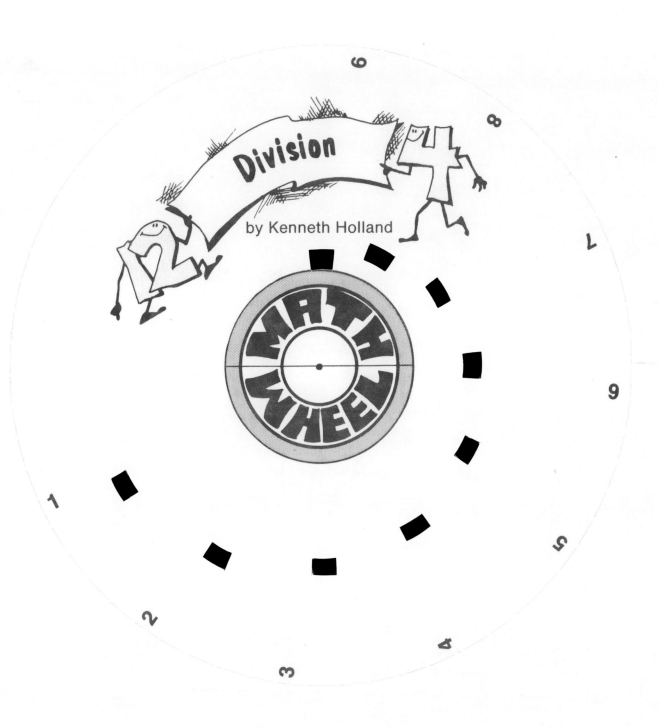

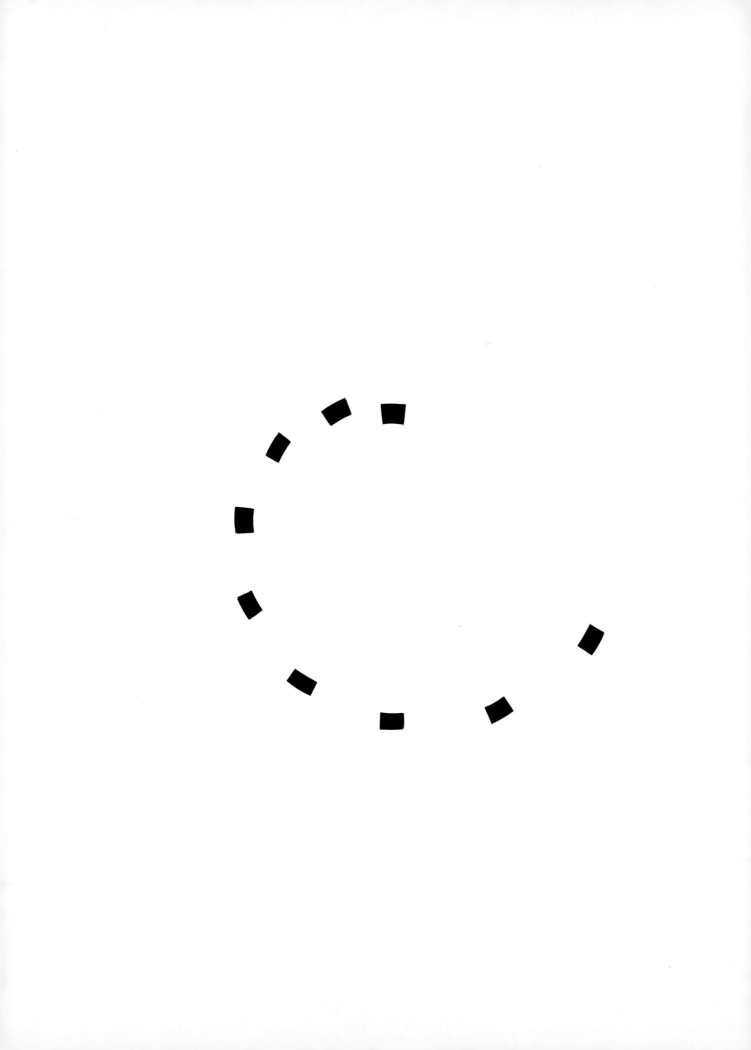

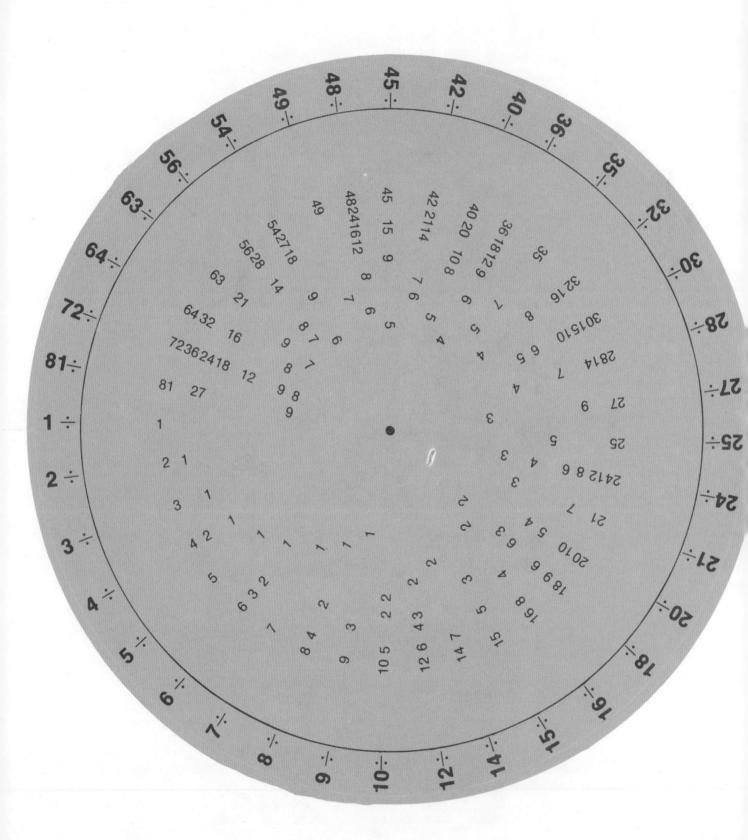